MILLMARK EDUCATION

ConceptLinks™

LITERACY AND LANGUAGE THROUGH CONTENT

Stars and Galaxies

Teacher's Guide

Millmark
EDUCATION

Program Authors

Mary Hawley
Program Author, Instructional Design

A former classroom teacher, Mary Hawley has more than 20 years of experience developing successful K-12 products in language development, literacy, and the content areas.

Kate Boehm Jerome
Program Author, Science

Kate Boehm Jerome has authored over 80 science books for children. Her programs have won three NSTA/CBC Outstanding Science Trade Books Awards and a Teachers' Choice Award.

Program Advisors

Scott K. Baker, PhD
Pacific Institutes for Research
Eugene, OR

Carla C. Johnson, EdD
University of Toledo
Toledo, OH

Donna Ogle, EdD
National-Louis University
Chicago, IL

Betty Ansin Smallwood, PhD
Center for Applied Linguistics
Washington, D.C.

Gail Thompson, PhD
Claremont Graduate University
Claremont, CA

Emma Violand-Sánchez, EdD
Arlington Public Schools
Arlington, VA (retired)

Program Reviewers

Candy Carro
William Floyd School District
Mastic Beach, NY

Paige Jerome
St. Lucie County Schools
Pt. St. Lucie, FL

Renee Mackin
Chicago Public Schools
Chicago, IL

Sheryl Powell
Kings Canyon Unified
School District
Reedley, CA

Cheryl Wolfel, EdD
Community Consolidated School
District 15
Palatine, IL

Content Reviewer

Tom Nolan
Operations Engineer
NASA Jet Propulsion Laboratory
Pasadena, CA

ConceptLinks™

LITERACY AND LANGUAGE THROUGH CONTENT

Table of Contents

PLANNING TIME

Get to Know the Components 4

Plan Your Objectives 6

Plan the Lessons 8

Place Your Students 12

LESSONS

Preview

Introduce Stars and Galaxies 14

Prepare for Reading 16

Build Academic Vocabulary 18

Explore Chapter 1

Develop Strategic Learners 20

Guide Reading 22

Explore Chapter 2

Guide Reading 32

Explore Chapter 3

Guide Reading 42

Wrap Up

Review Concepts and Vocabulary 52

Assess and Present 54

Reinforce and Extend

Launch Projects 62

Assign Further Reading 64

Teach Concepts with Visual Mini-lessons 66

ADDITIONAL RESOURCES

Newcomer Lesson and Learning Masters 70

Writer's Workshop and Learning Masters 76

Writing Rubric 82

Oral Language Rubric 83

Answers to the Learning Masters 84

Get to Know the Components

Welcome to **ConceptLinks**™, a research-based, content-area literacy program that helps all students develop the skills, strategies, and content knowledge they need for academic success with grade-level materials. Before introducing the module to students, use pages 4-13 to learn about the components, plan your objectives, plan your lessons, and place your students in the appropriate **Student Books**.

Core Components

STUDENT BOOKS

Four leveled **Student Books** give all students access to core concepts related to stars and galaxies. The **Student Books** provide differentiated content and activities that allow students to enter at an appropriate level and progress toward more challenging concepts and vocabulary knowledge.

Key Features:

• Engaging texts by award-winning writers present standards-based science content.

• Striking visuals support carefully leveled text.
• Embedded skills and strategies align with high-stakes testing.
• Response pages promote language development and science process skills through interactive work.
• Writing activities help students synthesize the learning as they communicate academic concepts.

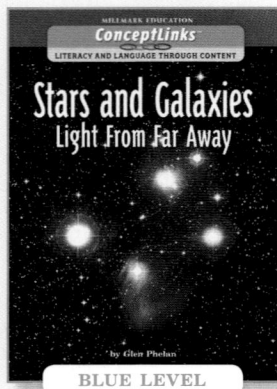

MILLMARK EDUCATION ConceptLinks LITERACY AND LANGUAGE THROUGH CONTENT **Stars and Galaxies** Light From Far Away by Glen Phelan BLUE LEVEL	MILLMARK EDUCATION ConceptLinks LITERACY AND LANGUAGE THROUGH CONTENT **Stars and Galaxies** Our Universe by Glen Phelan GREEN LEVEL	MILLMARK EDUCATION ConceptLinks LITERACY AND LANGUAGE THROUGH CONTENT **Stars and Galaxies** Changing Over Time by Glen Phelan ORANGE LEVEL	MILLMARK EDUCATION ConceptLinks LITERACY AND LANGUAGE THROUGH CONTENT **Stars and Galaxies** Exploring with Technology by Glen Phelan PURPLE LEVEL
Reading Level: Grade 2 **TESOL Level:** Starting Up/Emerging **Guided Reading:** L-M **Lexile™:** 500	**Reading Level:** Grade 3 **TESOL Level:** Developing **Guided Reading:** O-P **Lexile™:** 660	**Reading Level:** Grade 4 **TESOL Level:** Expanding **Guided Reading:** Q-R **Lexile™:** 740	**Reading Level:** Grade 5 **TESOL Level:** Bridging **Guided Reading:** U-V **Lexile™:** 790

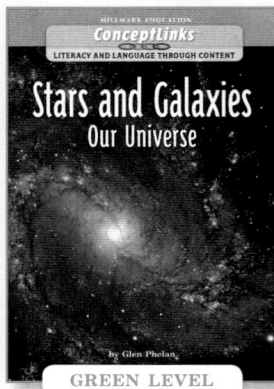

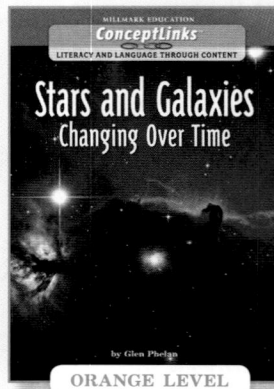

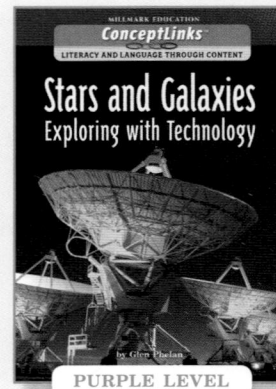

LINKS TO THE RESEARCH

"The acquisition of both language literacy and science literacy is dependent on the students' ability to think critically in similar ways."

– Judy McKee and Donna Ogle, *Integrating Instruction: Literacy and Science* (The Guilford Press, 2005)

CONCEPT CONNECTOR

The **Concept Connector** is a powerful visual tool that builds background, develops oral language, and informally assesses students' content knowledge.

Key Features:

- Content-related images help students access prior knowledge.
- Leveled questions reveal students' understanding of the core concepts.
- Vocabulary is presented and reviewed.

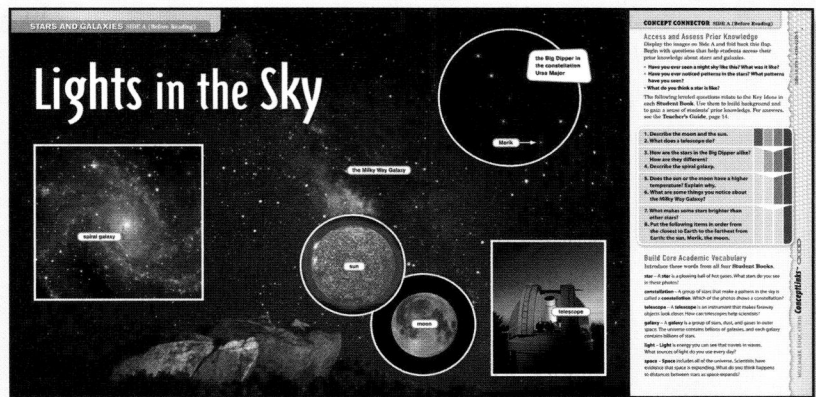

TEACHER'S GUIDE

This **Teacher's Guide** provides clear lesson plans with built-in differentiated instruction, adaptable for many classroom situations.

Key Features:

- Every lesson supports specific language, literacy, and content objectives.
- Step-by-step lessons make differentiated instruction meaningful and manageable.
- Optional activities reinforce and extend the learning.
- Formal and informal assessments measure student progress.

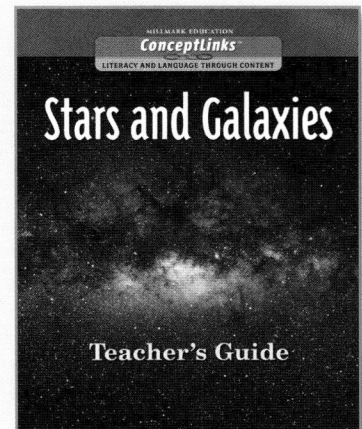

TECHNOLOGY

- The **Audio CD** provides oral readings of the **Student Books** and support for the Newcomer Lesson.
- The **Image Bank CD-ROM** offers downloadable photographs and visual aids that support vocabulary development, concept knowledge, writing activities, and student projects.
- The **Classroom Management and Assessment CD-ROM** provides both writeable and downloadable classroom management tools, such as lesson planners and small group management plans, as well as student assessment **Learning Masters** and **E-Masters**.
- The **Resource CD-ROM** includes downloadable versions of the **Teacher's Guide**, including the **Learning Masters**, as well as additional **E-Masters** including home language resources and graphic organizers.
- The *ConceptLinks*™ website at www.millmarkeducation.com/conceptlinks provides valuable resources, including links to careers websites.

Plan Your Objectives

In every *ConceptLinks*™ module, powerful literacy strategies, language skills, and science process skills are embedded in the instruction, along with standards-based science concepts. Use the charts that appear here to plan your instructional objectives. Also see specific language, literacy, and content objectives within the lessons.

Stars and Galaxies

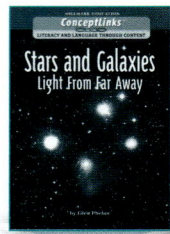

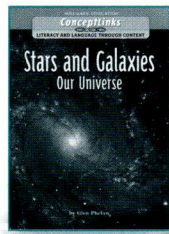

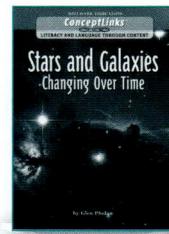

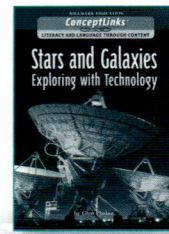

Stars and Galaxies — Light From Far Away
Stars and Galaxies — Our Universe
Stars and Galaxies — Changing Over Time
Stars and Galaxies — Exploring with Technology

TARGET SKILLS AND STRATEGIES	BLUE LEVEL	GREEN LEVEL	ORANGE LEVEL	PURPLE LEVEL
Comprehension Strategy	Ask Questions	Ask Questions	Ask Questions	Ask Questions
Science Process Skills	Observe Summarize Infer	Observe Summarize Infer	Observe Summarize Infer	Observe Summarize Infer
Speaking Focus	Ask Questions	Compare	Show Sequence	Explain
Writing Focus	Ask Questions	Compare	Show Sequence	Explain
Language Skill	Use *How* and *Why* in Questions	Words that Compare	Sequence Words	Explain by Comparing

LINKS TO NATIONAL STANDARDS

PreK-12 English Language Proficiency Standards (TESOL, 2006)

- Standard 1: English language learners communicate for social, intercultural, and instructional purposes within the school setting.

- Standard 2: English language learners communicate information, ideas, and concepts necessary for academic success in the area of language arts.

- Standard 4: English language learners communicate information, ideas, and concepts necessary for academic success in the area of science

National Science Education Standards (National Academy Press, 1996)

As a result of their activities, all students should develop:
- abilities necessary to do scientific inquiry (Content Standard A)
- understandings about scientific inquiry (Content Standard A)
- an understanding of the origin and evolution of the universe (Content Standard D)

Each *ConceptLinks*™ module provides differentiated, standards-based science content. The first level of the **Student Books**, the Blue Level, covers the most basic concepts related to the science topic. Each additional level covers the concepts of the previous level and develops additional concepts. This "spiraled" approach to grade-level content means that students can progress at a steady pace without being overwhelmed by new information.

The chart below shows how key concepts spiral through the Stars and Galaxies module.

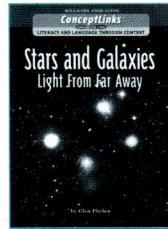

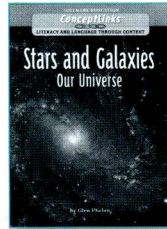

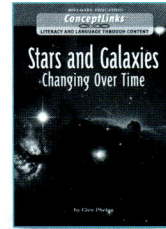

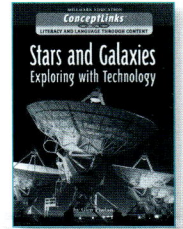

Stars and Galaxies

OVERVIEW OF KEY CONCEPTS	BLUE LEVEL	GREEN LEVEL	ORANGE LEVEL	PURPLE LEVEL
The light of bright objects in outer space comes from stars.				
Galaxies are huge groups of stars in space.				
Stars differ in terms of brightness, size, and color.				
Astronomers classify galaxies by different shapes.				
Stars have different life stages.				
Technology allows us to gather information on distant stars and galaxies in the universe.				

Place students in the **Student Books** that best match their instructional levels. (See the Placement Guide on pages 12-13.) Depending on your objectives and the amount of time you have available, use pages 8-11 of this guide to plan the reading of the **Student Books** and to select activities that will reinforce and extend the learning.

Plan the Lessons
(Two-week Pacing Plan)

These options are based on 40-60 minutes per day of instructional time. If you have less time or more time, use the suggestions here as a model for creating your own lesson plans. Also see the more detailed daily lesson plans on pages 10-11.

In the first five days...
Read and Respond to a Student Book

DAY 1	DAY 2	DAY 3	DAY 4	DAY 5
Preview	Explore Chapter 1 Your Turn Activities	Explore Chapter 2 Your Turn Activities	Explore Chapter 3 Your Turn Activities	Wrap Up Assess & Present

In the next five days...
Reinforce and Extend the Learning

	DAY 6	DAY 7	DAY 8	DAY 9	DAY 10
LAUNCH PROJECTS	Career Explorations Use Language/ Writer's Workshop	Career Explorations Writer's Workshop	Career Explorations Writer's Workshop	Science Around You Writer's Workshop	Wrap Up Assess & Present
AND/OR	AND/OR	AND/OR	AND/OR	AND/OR	AND/OR
ASSIGN FURTHER READING	Preview Another **Student Book**	Explore Chapter 1 Your Turn Activities	Explore Chapter 2 Your Turn Activities	Explore Chapter 3 Your Turn Activities	Wrap Up Assess & Present
AND/OR	AND/OR	AND/OR	AND/OR	AND/OR	AND/OR
TEACH VISUAL MINI-LESSONS	Throughout the five days, plan and teach visual mini-lessons as desired to introduce or reinforce the science concepts.				

Additional Pacing Plans

For one week of instruction:

- Have students read **Student Books** at their proficiency levels and respond to the Your Turn activities.
- If time permits, see Launch Projects (pages 62-63 in this guide) to choose Reinforce and Extend activities. You can also teach visual mini-lessons to introduce or reinforce the science concepts (pages 66-68).

For three weeks of instruction:

- **Week 1** Have students read **Student Books** at their proficiency levels and respond to the Your Turn activities.
- **Week 2** Choose Reinforce and Extend activities. See Launch Projects (pages 62-63) and Teach Concepts with Visual Mini-lessons (pages 66-68 of this guide).
- **Week 3** Have each student read another **Student Book** at a higher or lower level and respond to the Your Turn activities. See Assign Further Reading on pages 64-65 of this guide.

For four weeks of instruction:

- **Week 1** Have students read **Student Books** at their proficiency levels and respond to the Your Turn activities.
- **Week 2** Choose Reinforce and Extend activities. See Launch Projects (pages 62-63) and Teach Concepts with Visual Mini-lessons (pages 66-68 of this guide).
- **Week 3** Have each student read another **Student Book** at a higher or lower level and respond to the Your Turn activities. See Assign Further Reading on pages 64-65 of this guide.
- **Week 4** Choose Reinforce and Extend activities. See Launch Projects (pages 62-63) and Teach Concepts with Visual Mini-lessons (pages 66-68 of this guide).

Starting Up with Newcomers

ConceptLinks™ also provides 5 or 10 days of instruction (20 minutes per day) for students with little or no proficiency in English. See pages 70-75 of this guide.

DAY 1	DAY 2	DAY 3	DAY 4	DAY 5
Introduce Vocabulary	Build Concepts & Vocabulary	Write About Stars and Galaxies	Explore the **Student Book**	Build a Story

DAY 6	DAY 7	DAY 8	DAY 9	DAY 10
Preview the Blue Level **Student Book**	Explore Chapter 1	Explore Chapter 2	Explore Chapter 3	Wrap Up Assess & Present

Plan the Lessons

Daily Lessons at a Glance

Use these lesson summaries to plan your daily lessons.
All page numbers refer to this **Teacher's Guide**.

DAY 1	DAY 2	DAY 3
PREVIEW	**EXPLORE CHAPTER 1**	**EXPLORE CHAPTER 2**
Introduce Stars and Galaxies with the **Concept Connector**, Side A 15-20 minutes (pages 14-15) • Access and Assess Prior Knowledge (whole group) • Build Core Academic Vocabulary (whole group) **Prepare for Reading** 15-20 minutes (pages 16-17) • Preview the **Student Book** (small groups) • Develop Language (small groups) **Build Academic Vocabulary** 10-15 minutes (pages 18-19) • Generate a Word List (whole group) • Create a Word Map (whole group)	**Develop Strategic Learners** 15-20 minutes (pages 20-21) • Teach the Comprehension Strategy (whole group) • Set Purposes for Reading (partners) **Guide Reading** 25-40 minutes (pages 22-31) • Review Concepts (whole group) • Teach the Mini-lesson: Observe (whole group) • Read Chapter 1 (small groups) • Respond and Reflect: Your Turn (small groups)	**Guide Reading** 25-40 minutes (pages 32-41) • Review Concepts (whole group) • Teach the Mini-lesson: Summarize (whole group) • Read Chapter 2 (small groups) • Respond and Reflect: Your Turn (small groups)
	Reinforce and Extend **Launch Projects** Career Explorations 10-45 minutes (page 62) partners, small groups	**Reinforce and Extend** **Launch Projects** Use Language/Writer's Workshop 20-120 minutes (page 63, pages 76-81) small groups, partners, individuals

DAY 4	DAY 5	DAYS 6-10
EXPLORE CHAPTER 3	**WRAP UP**	**REINFORCE AND EXTEND** (Optional Activities)

Guide Reading
25-40 minutes
(pages 42-51)

- Review Concepts
  (whole group)

- Teach the Mini-lesson: Infer
  (whole group)

- Read Chapter 3
  (small groups)

- Respond and Reflect: Your Turn
  (small groups)

Review Concepts and Vocabulary with the **Concept Connector**, Side B
15-20 minutes
(pages 52-53)

- Review Concepts
  (whole group)

- Review Key Words
  (whole group)

Assess and Present
30-60 minutes
(pages 54-61)

- Administer Written Assessments
  (individuals)

- Observe Oral Language
  (individuals, partners, small groups)

- Guide Presentations
  (individuals, partners, whole group)

Launch Projects
5-180 minutes
(pages 62-63)

(See references to Career Explorations, Use Language/Writer's Workshop, and Science Around You on Days 2-5. These projects can be completed during Days 2-5 or Days 6-10.)

Assign Further Reading
30-60 minutes
(pages 64-65)

- Select Another Student Book
  (teacher planning time)

- Read Another Student Book
  (individuals, partners, small groups)

Teach Concepts with Visual Mini-lessons
5-15 minutes per mini-lesson
(pages 66-68)

- Plan a Visual Mini-lesson
  (teacher planning time)

- Teach the Mini-lesson
  (whole group, small groups)

Reinforce and Extend
Launch Projects
Science Around You
5-15 minutes
(page 63)
partners, individuals

Writer's Workshop (cont'd)

Reinforce and Extend
Launch Projects
Writer's Workshop (cont'd)

Place Your Students

Use the Placement Guide and your own observations to place each student in the **Student Book** that best matches his or her instructional reading and language levels.

Placement Guide

READING LEVEL / TESOL LEVEL	GRADE 1 OR BELOW NEWCOMER/ STARTING	GRADES 1-2 EMERGING
READING COMPREHENSION	**If the student...** • constructs meaning primarily from non-text sources	**If the student...** • reads words and phrases • locates specific information in print
LANGUAGE PROFICIENCY	**If the student...** • responds nonverbally to simple commands, statements, and questions • has a use of English that is limited to a few words, phrases, and chunks of language • has a very limited comprehension of oral English in both social and academic contexts	**If the student...** • has a use of English that is limited to high-frequency words, some general academic vocabulary, common expressions, and short sentences • has some comprehension of oral English in social contexts • has a very limited comprehension of oral English in academic contexts
WRITING PROFICIENCY	**If the student...** • uses words and phrases to generate simple texts • generates texts that show nonconventional features (such as invented spelling, grammatical errors, features of home language)	**If the student...** • writes words, phrases, and short sentences but with errors that hinder comprehension

Then place the student as follows:

STARTING UP WITH NEWCOMERS
Newcomer Lesson
pages 70-75

BLUE LEVEL
Stars and Galaxies: Light from Far Away

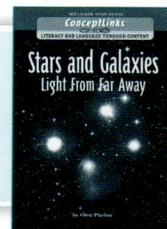

NOTE: Student profile descriptions are adapted from the *TESOL PreK-12 English Language Proficiency Standards,* TESOL 2006, pages 39-41.

GRADES 3-4 DEVELOPING	GRADES 4-5 EXPANDING	GRADES 5-6 BRIDGING
If the student... • with prior knowledge is able to construct meaning from text	**If the student...** • struggles at times with complex structures or abstract vocabulary	**If the student...** • reads fluently but benefits from some modifications of grade-level materials
If the student... • uses fairly fluent English but may have difficulty expressing ideas due to limited vocabulary and lack of knowledge of language structures • comprehends oral English in social contexts • has some comprehension of oral English in academic contexts	**If the student...** • uses fluent English and knows some specialized academic vocabulary • has a strong comprehension of oral English in social contexts • has a good comprehension of oral English in academic contexts but lacks specialized vocabulary	**If the student...** • uses fluent English in social and academic contexts, with good command of academic vocabulary • still has some difficulties with grade-level academic discourse, or gaps in content knowledge
If the student... • generates texts that are more complex and more coherent • generates texts that still have frequent errors and nonconventional features	**If the student...** • generates texts that begin to approximate those of native speakers of English • generates texts with errors that persist but do not hinder communication	**If the student...** • generates texts that approximate those of native speakers of English • generates texts with minimal errors

Then place the student as follows:

GREEN LEVEL
Stars and Galaxies: Our Universe

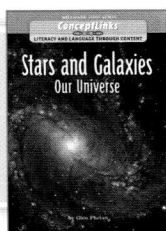

ORANGE LEVEL
Stars and Galaxies: Changing Over Time

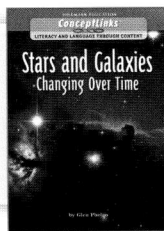

PURPLE LEVEL
Stars and Galaxies: Exploring with Technology

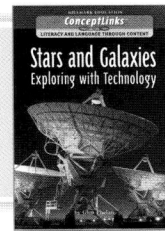

PREVIEW

LESSON ACTIVITIES

Introduce Stars and Galaxies
15-20 minutes (TG, pp. 14-15)
- Access and Assess Prior Knowledge
- Build Core Academic Vocabulary

Prepare for Reading
15-20 minutes (TG, pp. 16-17)
- Preview the Student Books
- Develop Language

Build Academic Vocabulary
10-15 minutes (TG, pp. 18-19)
- Generate a Word List
- Create a Word Map

OBJECTIVES

Students will:

LANGUAGE
USE academic language to communicate information, ideas, and concepts in science.

LITERACY
ACCESS prior knowledge to prepare for reading.

CONTENT
UNDERSTAND that the lights we see in the night sky come from billions of stars.

MATERIALS

- **Concept Connector**, Side A
- **Learning Master 1**
- **Image Bank CD-ROM**
- **Resource CD-ROM**

ResourceLinks

See **E-Masters 4**, **4A**, **5**, and **5A** on the **Resource CD-ROM**.

Introduce Stars and Galaxies

Activity Overview

Access and Assess Prior Knowledge	whole group	10-15 minutes
Build Core Academic Vocabulary	whole group	5 minutes

Access and Assess Prior Knowledge

Distribute copies of the Concept Guide, **Learning Master 1**. Display the **Concept Connector**, Side A, folding back the Side A flap. You can also use the **Image Bank CD-ROM** to print or project the Side A images. Use the teaching notes on the Side A flap to guide a class discussion. Students can use the Concept Guide to label the images and record main points from the discussion. For the leveled questions, build background by providing these answers if students are unable to do so:

Answers for Concept Connector, Side A

1. Students might include that the sun makes its own light and is very bright, and the moon does not make its own light and is duller than the sun.

2. A telescope makes faraway objects look bigger and closer.

3. Students might include that some of the stars appear to be the same color or brightness, and some differ in brightness, color, and size.

4. Possible response: The spiral galaxy has many stars. Its shape looks like a pinwheel.

5. Students might say that the sun is hotter than the moon because the sun produces heat and the moon does not.

6. Possible response: Some stars appear to be brighter or a different color than others, and there are so many stars in the Milky Way Galaxy that the sky looks milky.

7. Students may know that one star may be brighter than others because it is larger, hotter, or closer to Earth.

8. The correct order from closest to Earth to farthest from Earth is: the moon, the sun, Merik.

Build Core Academic Vocabulary

Use the scripted notes on the Side A flap to introduce the core academic vocabulary for the four Stars and Galaxies **Student Books**. Continue discussing the images on the **Concept Connector**, Side A, incorporating these words.

CORE ACADEMIC VOCABULARY

star	telescope	light
constellation	galaxy	space

NAME

Lights in the Sky

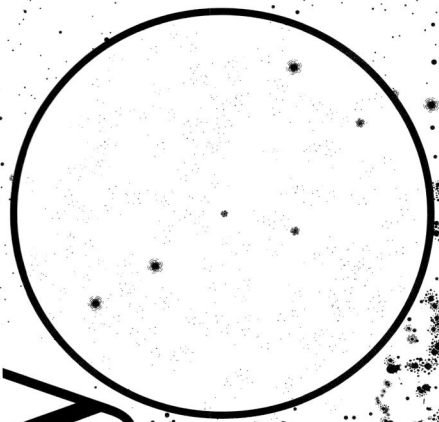

the Milky Way Galaxy

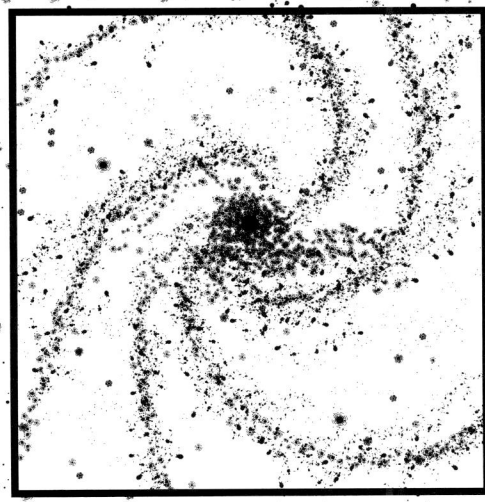

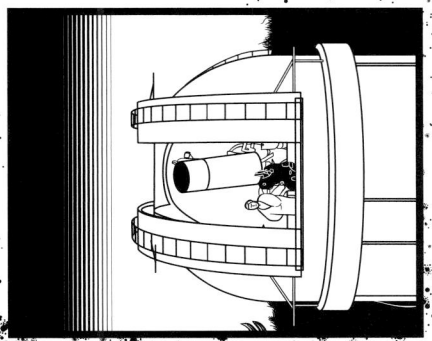

ALL LEVELS

OBJECTIVES

Students will:

LANGUAGE

USE appropriate discourse patterns to discuss stars and galaxies.

LITERACY

PREVIEW text and **MAKE PREDICTIONS** to prepare for reading.

CONTENT

DESCRIBE how stars and galaxies appear in the night sky.

MATERIALS

- **Student Books**, all levels
- **Learning Master 2**

ResourceLinks

Allow students to use the Search section of the **Image Bank** to find visual support for new or difficult words.

Prepare for Reading

Activity Overview

Preview the Student Books	small groups	
Develop Language	small groups	15-20 minutes

Preview the Student Books

Give each student a copy of the Stars and Galaxies **Student Book** that you have assigned to him or her, as well as the Preview Guide, **Learning Master 2**. Form groups of students reading the same title.

- Read aloud the instructions on the Preview Guide. Point out key features in the **Student Books** such as the inside front cover, the table of contents, the chapters and Your Turn pages, and the feature pages at the end.
- Have students work together in their small groups to complete the Preview Guide. While students are engaged in this task, visit each group and lead the Develop Language activity on pages 2-3 of each **Student Book**.

Develop Language

As you visit each small group, have students turn to the Develop Language feature on pages 2-3 in their **Student Book**. Invite students to share any prior knowledge they have about the stars, galaxies, and telescopes shown in the photos. Next, use the sample questions and answers to lead an instructional conversation about the objects in the night sky and the telescopes that we use to study them. Encourage students to ask and answer other questions about what telescopes might help us learn about the stars.

	BLUE LEVEL	GREEN LEVEL	ORANGE LEVEL	PURPLE LEVEL
FEATURED SETTING	Star clusters	The Milky Way Galaxy	Stars in the night sky	Observatories and telescopes
QUESTION AND ANSWER PATTERNS	What do you see in the sky? The sky is full of ____ . What do the stars look like? They look like ____ . Do you see any...? I see a ____ . What questions do you have...?	Describe the sun. The sun is ____ . It is also ____ . What do the other images show...? The other images show that ____ . Which image shows...? What questions...?	Which photo was...? The longer exposure was used to make photo ____ . How might...? This part of the sky would ____ .	How do these ____ differ in ____ ? How do they differ in ____ ? In what other ways...?

Book Title _____   ALL LEVELS

Look at the inside front cover of your book.
Read the Science Vocabulary Words. Which words do you already know?
Write each word in one of the columns below.

SCIENCE VOCABULARY WORDS

I know these words well.	I know these words a little.	I do not know these words.

Look through your book. Look at the pictures. Read the labels and captions.
Write your ideas about each chapter on the chart.

	Name things you see in the pictures.	List what you wonder about the chapter.	Ask a question about an interesting picture.
Chapter 1 Title:		I wonder	
Chapter 2 Title:		I wonder	
Chapter 3 Title:		I wonder	

ALL LEVELS

OBJECTIVES

Students will:

LANGUAGE
IDENTIFY, **EXPLAIN**, and **ELABORATE** to explore concepts.

LITERACY
USE graphic organizers to expand vocabulary.

CONTENT
DEVELOP an understanding of key words related to stars and galaxies.

MATERIALS

- **Student Books**, all levels
- **Learning Master 3**
- **Resource CD-ROM**

HOME LANGUAGE SUPPORT

Invite students to investigate whether any of the suggested words are cognates for home language words. For home language support, see the **Resource CD-ROM**.

Build Academic Vocabulary

Activity Overview

| Generate a Word List | whole group | 5 minutes |
| Create a Word Map | whole group | 5-10 minutes |

Generate a Word List

Distribute copies of the Word Map, **Learning Master 3**. Display a larger version of the Word Map on chart paper, on a transparency, or on the board.

- Point out the Word List. Ask: **What are some other words about stars and galaxies we might find in our reading?** Invite students to look through their **Student Books** to suggest other words.
- Add the suggested words to the larger version of the Word List as students record them on **Learning Master 3**. Discuss the words, exploring their meanings in ways that allow students at all proficiency levels to participate.

Provide Scaffolding for English Language Learners

STARTING/EMERGING	DEVELOPING	EXPANDING	BRIDGING
When a word is suggested, have students repeat the word and count the syllables.	When a word is suggested, have students ask each other questions about the word.	When a word is suggested, have students create and discuss a definition of the word.	When a word is suggested, have students give an example of the word or use it in a sentence.

Create a Word Map

Begin a Word Map like the one below. First, have students discuss how the words are related to the target word, **galaxy**. Then work with students to map the words. If students don't know enough about a word yet to map it, it can be added after reading. Have students save their maps for later reference.

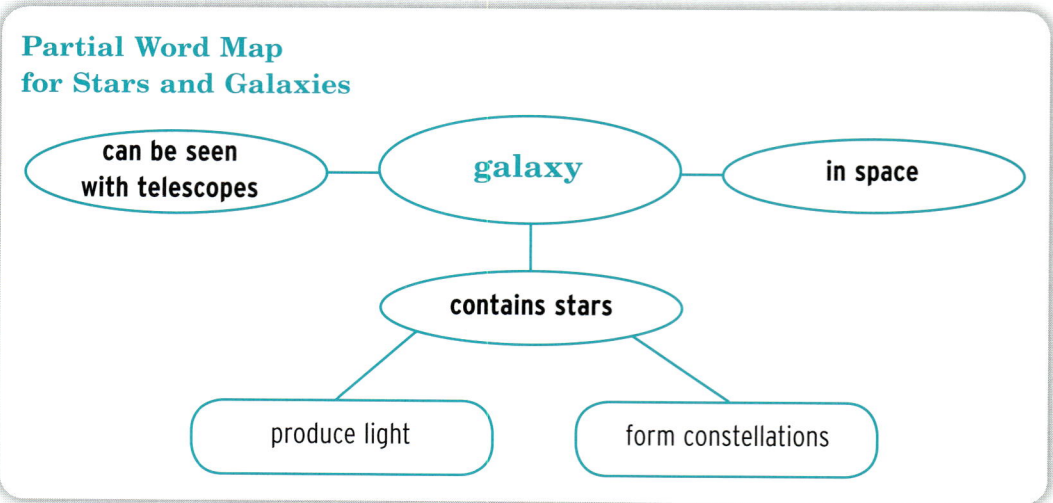

Partial Word Map for Stars and Galaxies

can be seen with telescopes — galaxy — in space

contains stars

produce light   form constellations

Lesson Wrap-up Invite students to share their questions and predictions about the books they will read.

ALL LEVELS

NAME ____________________

Explore Vocabulary

Look at the Word List. Add other words about galaxies. Talk about all the words. How do they relate to galaxies? Make a Word Map. Show how these words about stars and galaxies are connected.

galaxy

WORD LIST

constellation
light
space
star
telescope

ALL LEVELS

LESSON ACTIVITIES

Develop Strategic Learners
15-20 minutes (TG, pp. 20-21)
- Teach the Comprehension Strategy
- Set Purposes for Reading

Guide Reading
25-40 minutes (TG, pp. 22-31)
- Review Concepts
- Mini-lesson: Observe
- Read Chapter 1
- Respond and Reflect
- Reinforce and Extend

OBJECTIVES

Students will:

LANGUAGE

USE academic language to discuss comprehension strategies.

LITERACY

LEARN and **APPLY** the comprehension strategy of asking questions.

SET purposes for reading.

CONTENT

UNDERSTAND that most stars are binary stars.

MATERIALS

- **Learning Master 4**
- **Audio CD**, Tracks 6, 10, 14, 18
- **Resource CD-ROM**

ResourceLinks

For a visual comprehension strategy lesson, see **E-Master 8** on the **Resource CD-ROM**. Also see Support for Key Ideas on the **Image Bank**.

Develop Strategic Learners

Activity Overview

Teach the Comprehension Strategy	whole group	10-15 minutes
Set Purposes for Reading	partners	5 minutes

Teach the Comprehension Strategy: Ask Questions

INTRODUCE Direct students' attention to the diagram. Explain that our solar system has only one star, the sun, but many star systems have two stars. Have students ask questions about the diagram and share their questions. Tell students that good readers ask questions before, during, and after they read.

MODEL Read aloud the article. Then say: **As I read this article, I think of questions: Why do these stars orbit each other? Would they ever crash into each other? What would a sunset in a solar system with two suns look like? Good readers pay attention to what questions they have about the content and what they wonder as they read. Then they find answers to their questions by rereading the text or using other sources of information.** Have students use the text to locate answers to some of their questions. Ask: **Did you have questions that the text did not answer? Where would you look for those answers?**

PRACTICE/APPLY Have students work together to fill in the chart and answer the questions at the bottom of the Strategy Guide, **Learning Master 4**.

Set Purposes for Reading

Pair students who are reading the same levels of the **Student Books**. Have them refer to the inside front cover of their books to do the following:

- **Review** the vocabulary words, noting any words not yet discussed in class.
- **Read aloud** the Strategy Focus to recap the comprehension strategy.
- **Read** and **think** about language that **asks questions** (BLUE LEVEL), **compares** (GREEN LEVEL), **shows sequence** (ORANGE LEVEL), or **explains** (PURPLE LEVEL).
- **Set purposes** by reading the Set a Purpose for Reading statement, discussing what they already know about the topic, and adding questions they have about stars and galaxies.

For extra support and pronunciation practice, have students listen to the appropriate track on the **Audio CD**.

Ask Questions

Binary Stars

Our solar system has one star: the sun. But most stars are **binary stars**, or stars that have formed in pairs.

Binary stars form close together in a cloud of gas and dust. The gravity of each star pulls on the other star. This pull causes the stars to orbit each other. Some stars are very close together. They look like a single star even through a telescope.

Binary stars orbit around a common point called a **center of gravity**. The time it takes to complete an orbit varies greatly. It may be as short as thirty minutes, or as long as millions of years. The time depends on how close the stars are to each other.

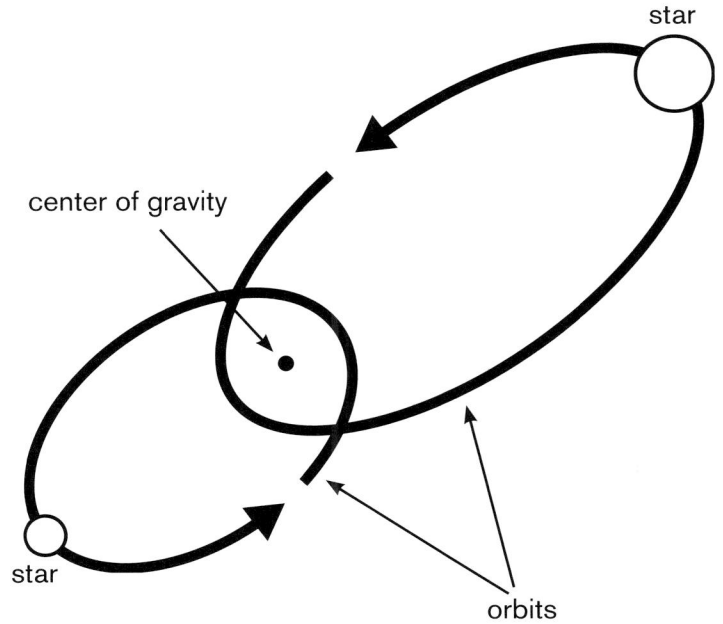

star

center of gravity

star

orbits

Binary Star System

STRATEGY FOCUS: ASK QUESTIONS

1. Look at the article title and diagram on this page. Before you read the article, write one question you have about binary stars. As you read the article, write another question. Then write one question you still have after reading the article.

Before Reading	During Reading	After Reading

2. After reading the article, reread your questions. Write any answers you found. How can you find answers to the questions you still have?

_____

_____

_____

OBJECTIVES

Students will:

LANGUAGE

USE academic language to ask and answer questions about stars and galaxies.

LITERACY

MAKE CONNECTIONS from text to self and from text to world.

CONTENT

APPLY the science process skill of observing.

See the book-level Guide Reading lessons for additional objectives.

MATERIALS

- **Student Books**, all levels
- **Learning Masters 5-8**
- **Audio CD**, Tracks 7, 11, 15, 19
- **Resource CD-ROM**
- **Image Bank CD-ROM**

ResourceLinks

Use **E-Master 3** on the **Resource CD-ROM** to help manage small-group activities. See the **Image Bank**, Support for Key Ideas, for downloadable versions of maps, charts, or diagrams in the chapter.

Guide Reading

Activity Overview

Review Concepts	whole group	2-3 minutes
Teach the Mini-lesson: Observe	whole group	5 minutes
Read Chapter 1 Respond and Reflect	small groups small groups	20-30 minutes
Reinforce and Extend	small groups, individuals	10-45 minutes (optional)

Review Concepts

Have students review their Word Maps (**Learning Master 3**). Ask them to share what the Word Maps show about stars and galaxies.

Teach the Mini-lesson

Science Process Skill: Observe

SHOW STUDENTS a close-up color photo of a star. Have them observe and identify different properties of the star. List the properties on the board as students name them.

TELL STUDENTS When you observe, you look carefully at an object. You ask questions such as: What is this star made of? What shape is it? What color is it? How bright is it? Scientists learn about stars by making careful and accurate observations. They use special instruments, such as special cameras and telescopes to see more stars, see stars more clearly, and to make images of stars.

TELL STUDENTS that they will be making observations about photos of galaxies and stars.

Read Chapter 1

Use this grouping plan as a guide for planning reading time.

BLUE LEVEL	GREEN LEVEL	ORANGE LEVEL	PURPLE LEVEL
Students read with your guidance.	Students read with a partner or with the **Audio CD**, Track 11.	Students read the chapter alone or with a partner.	Students read the chapter alone or with a partner.
Students reread with a partner or with the **Audio CD**, Track 7.	**Students reread with your guidance.**	Students work together to complete the Study Guide, LM 7.	Students work together to complete the Study Guide, LM 8.
Students work together to complete the Study Guide, LM 5.	Students work together to complete the Study Guide, LM 6.	**Students reread with your guidance.**	Students begin the Your Turn activities, SB page 9.
Students begin the Your Turn activities, SB page 9.	Students begin the Your Turn activities, SB page 9.	Students begin the Your Turn activities, SB page 9.	**Students review with your guidance.**

LM = Learning Master   SB = Student Book

- Before students read, distribute the appropriate Study Guides and verify that students understand the tasks.
- As students read, circulate among the small groups to guide the reading and monitor understanding. Use the Guide Reading lessons on pages 24, 26, 28, and 30. Have students complete the book-level Study Guide.
- After students read, have them work in their small groups or with a partner to complete the Your Turn activities on page 9 of the **Student Books**.

Respond and Reflect: Your Turn

Students respond to the reading with hands-on activities that include applying the science process skill, making connections, and building oral fluency with academic language.

HOME LANGUAGE SUPPORT

Students may engage more deeply in the response activities if their initial discussions are in their home language.

	BLUE LEVEL	GREEN LEVEL	ORANGE LEVEL	PURPLE LEVEL
OBSERVE	Students observe the stars in a photo and tell how they look different. **Possible response:** Some stars look bright and some stars look dim. Most stars look white, but some stars look red or blue.	Students shine a flashlight on different objects and observe what each object looks like. **Possible response:** 1. Both give off light. 2. Some objects seem to shine brighter because they have shiny surfaces or are closer to the flashlight. 3. Most of the objects I see do not produce light. Some objects, such as the light in the fish tank, do produce their own light.	Students observe a photo of two stars and answer questions that compare them. **Correct response:** 1. The blue star has a hotter temperature because blue stars are hotter than red stars. 2. The blue star has a greater apparent magnitude because it appears brighter than the red star.	Students observe photos of telescopes and tell which telescopes collect something other than light. **Possible response:** 1. I think the Arecibo telescope collects something other than light. 2. It looks different from the other telescopes and its name says it is a radio telescope.
MAKE CONNECTIONS	Students consider where they can see the stars at night and tell a friend about what they can see. **Possible response:** I can see stars in the sky away from city lights. There are bright and dim stars in the night sky. Some stars look very close together.	Students calculate the distance from Earth to Arcturus in kilometers. **Correct response:** Arcturus is 342 trillion kilometers from Earth.	Students compare the apparent magnitude and absolute magnitude of the sun and the star Rigel. **Correct response:** The sun has a greater apparent magnitude, but Rigel has a greater absolute magnitude.	Students name objects in space that shine because they reflect starlight. **Possible response:** Objects in space that reflect starlight include the moon, planets, and satellites.
USE THE LANGUAGE OF SCIENCE	Have students practice a fluent use of academic language by asking and answering the question with a partner. Then invite them to compose and practice other questions and answers about the chapter.			

Reinforce and Extend: Career Explorations (optional)

If you have a longer instructional period, students can carry out the Career Explorations activity on page 20 of the **Student Books**. See the teaching notes on page 62 of this guide.

ResourceLinks

Allow students to search the **Image Bank** for photos relating to the featured careers.

Lesson Wrap-up Invite students to share their responses to the Your Turn activities with the whole class or in small groups.

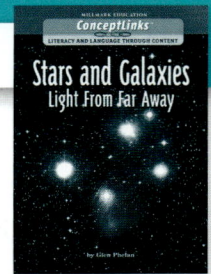

Stars and Galaxies
Light From Far Away

OBJECTIVES

Students will:

LANGUAGE

USE knowledge of Latin words to determine meaning.

LITERACY

READ and **ANALYZE** nonfiction texts.

USE visuals to aid comprehension.

ASK questions to deepen understanding.

CONTENT

UNDERSTAND that a constellation is a group of stars that makes a pattern.

UNDERSTAND that telescopes help scientists learn about faraway stars.

MATERIALS

• **Student Book**, Blue Level
• **Learning Master 5**
• **Audio CD**, Track 7

FOR MORE SUPPORT

PAGE 5 Explain how to decode a long word, such as **constellation**, by dividing it into syllables: con / stel / la / tion. Model how to sound out each syllable. Then help students find the words **telescope** (page 6) and **scientists** (page 7) in the chapter and repeat the exercise.

Guide Reading

Activity Overview

Blue Level	small group	5-10 minutes

• Use these leveled notes to guide the reading of Chapter 1 in the Blue Level **Student Book**.
• Have the other book-level small groups work on their own Chapter 1 activities. See the detailed grouping plan on page 22.
• When you leave the group, remind students to complete **Learning Master 5** and the Your Turn activities on page 9 of the **Student Book**.
• Return to page 23 of this guide for answers to the Your Turn activities and additional teaching suggestions.

Read Chapter 1: Looking at Stars

Page 4 Read the title. Ask: **What do you see in the photo?** (stars) **What do you observe about them?** (Some look brighter and some look larger than others.) Hold up the page and trace the pattern of the Big Dipper with your finger. Ask: **Can you name the shape that this pattern forms?** (Answers will vary.) Draw the pattern on the board and connect the stars to show the Big Dipper.

Page 5 Ask students to look at the image. Read aloud the Key Idea. Write **constellation** on the board. Ask: **What is a star pattern called?** (a constellation) Have students find the Big Dipper in the constellation Ursa Major.

> **EXPLORE LANGUAGE** Explain that in some languages, such as Latin, Spanish, and French, adjectives often follow the noun they modify. Explain that in English, adjectives almost always come before the nouns they modify.

Pages 6-7 Have students look at the photos and read the captions. Ask: **What else can you see in the night sky with a telescope?** (Possible response:

the moon and planets) Ask: **How is the telescope on page 7 different from the one on page 6?** (It has a different shape. It is larger. It can be used to see faraway stars more clearly.)

Page 8 Ask students to observe the location of the Hubble telescope. Ask: **How is the location of the Hubble telescope different from the locations of the telescopes on pages 6 and 7?** (The Hubble telescope orbits Earth. The other two telescopes are on Earth's surface.) Explain that the Hubble telescope takes pictures of stars. Have students look at the photo in the circle and describe what they see.

Apply the Comprehension Strategy

ASK QUESTIONS Remind students that good readers ask questions to learn more. Ask: **What questions do you have about the photos on page 8?** (Possible response: How far away are the stars? Have we ever used the Hubble telescope to discover a new star?)

Study Guide
Learning Master 5

Stars and Galaxies: Light from Far Away
Chapter 1: Looking at Stars

BLUE LEVEL
Student Book,
pages 4-8

USE KEY WORDS

Look at the Key Words on page 23 of your book.
Answer these questions about the Key Words in Chapter 1.

KEY WORDS
constellation
stars
telescopes

1. A **constellation** is a group of _____ . Select the best answer.

   A. stars   **B.** telescopes   **C.** planets   **D.** buildings

2. The _____ Ursa Major looks like a bear.

3. What do scientists use to look into space?

   _____

ORGANIZE IDEAS

As you read Chapter 1, complete the concept map.

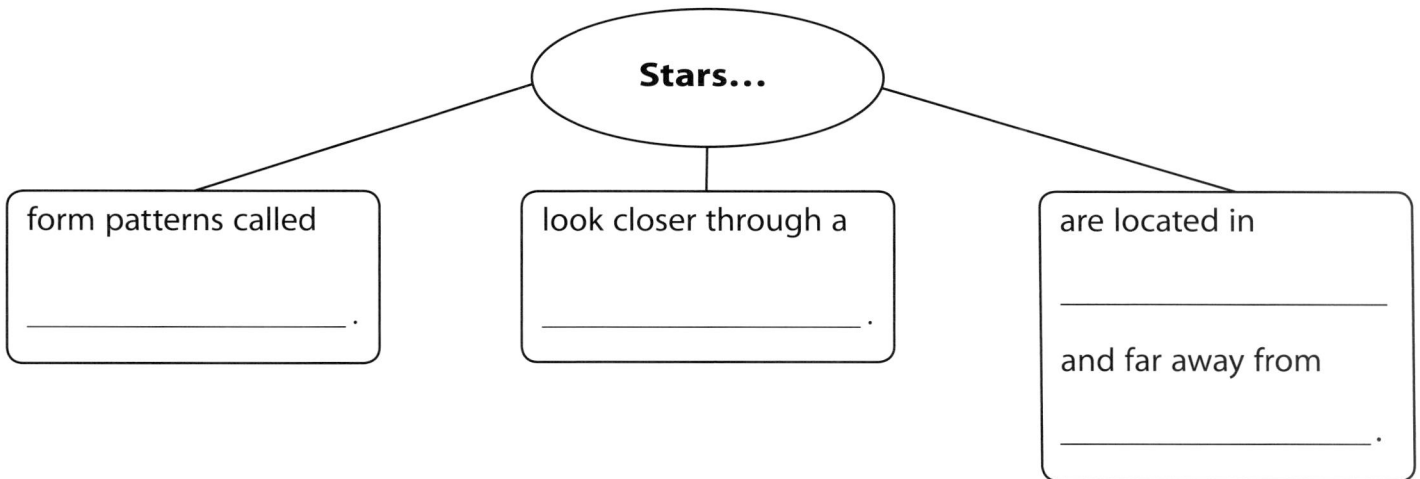

Stars...

form patterns called

_____ .

look closer through a

_____ .

are located in

_____

and far away from

_____ .

STRATEGY FOCUS: ASK QUESTIONS

What is one question you still have about stars?

_____

_____

How could you find the answer to your question?

_____

_____

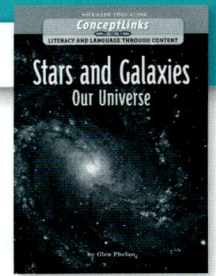

EXPLORE CHAPTER 1 (cont'd)

GREEN LEVEL
pages 4-8

ConceptLinks
LITERACY AND LANGUAGE THROUGH CONTENT
Stars and Galaxies
Our Universe
by Glen Phelan

OBJECTIVES

Students will:

LANGUAGE

SPEAK and **LISTEN** to share ideas about the sun and other stars.

LITERACY

READ and **ANALYZE** nonfiction texts.

INTERPRET diagrams.

ASK questions to deepen understanding.

CONTENT

UNDERSTAND that stars make their own light because atoms combine and release energy.

UNDERSTAND that a light-year is a measure of distance to stars.

MATERIALS

• **Student Book**, Green Level
• **Learning Master 6**
• **Audio CD**, Track 11

FOR MORE SUPPORT

PAGE 8 Read aloud the first sentence. Explain that the words **not all** mean "some." Some bright lights in space are stars, but some are not stars.

Guide Reading

Activity Overview

Green Level	small group	5-10 minutes

• Use these leveled notes to guide the reading of Chapter 1 in the Green Level **Student Book**.
• Have the other book-level small groups work on their own Chapter 1 activities. See the detailed grouping plan on page 22.
• When you leave the group, remind students to complete **Learning Master 6** and the Your Turn activities on page 9 of the **Student Book**.
• Return to page 23 of this guide for answers to the Your Turn activities and additional teaching suggestions.

Read Chapter 1: The Sun and Other Stars

Pages 4-5 Read aloud the chapter title. Have students describe what they observe in the photo of stars. Ask: **What is each point of light in the night sky?** (a star) Read aloud the definition of **atoms** on page 5. Explain that in this book, important words will be in bold when they appear for the first time and that these words will be defined on the page. Have a student read the Key Idea. Explain that everything students can see and feel is made of matter. Ask: **What forms of energy reach Earth from the sun?** (heat and light)

Pages 6-7 Read aloud the definition of **light-year** and the Key Idea. Direct students' attention to the diagram, and read aloud the caption. Ask: **How far does light travel in one light-year?** (9.5 trillion kilometers or 6 trillion miles) Have students compare the distance from Earth to the sun and from Earth to Proxima Centauri. Make sure students understand that the objects in the diagram are not to scale. Point out that Earth is actually much smaller than the sun and other stars. Earth is shown larger here to show that we are looking at the sun and other stars from Earth. The distances from Earth to the sun and to Proxima Centauri are also not to scale. If they were, Proxima Centauri would not fit on the diagram.

Page 8 Read aloud the definition of **reflect**. Read and discuss the captions and the Key Idea. Ask: **Which objects in space make light?** (stars) **Why does the moon look bright?** (It reflects light from the sun.) **What else in space reflects starlight?** (planets, gas, dust)

Apply the Comprehension Strategy

ASK QUESTIONS Draw attention to the photo on page 5. Remind students that asking questions as they read will help them to better understand the text. Ask: **What questions do you have about the sun?** (Possible response: Will the sun ever burn out?) Have students explain where they could find the answers to their questions.

Stars and Galaxies: Our Universe
Chapter 1: The Sun and Other Stars

GREEN LEVEL
Student Book,
pages 4-8

USE KEY WORDS

Look at the Key Words on page 23 of your book.
Answer these questions about the Key Words in Chapter 1.

KEY WORDS
light-year
star

1. A **light-year** is a measure of _____ .
   Select the best answer.
   A. brightness   **B.** distance   **C.** force   **D.** time

2. The **star** that is closest to Earth is _____ .

ORGANIZE IDEAS

As you read Chapter 1, complete the concept map to explain how energy forms in the sun and reaches Earth.

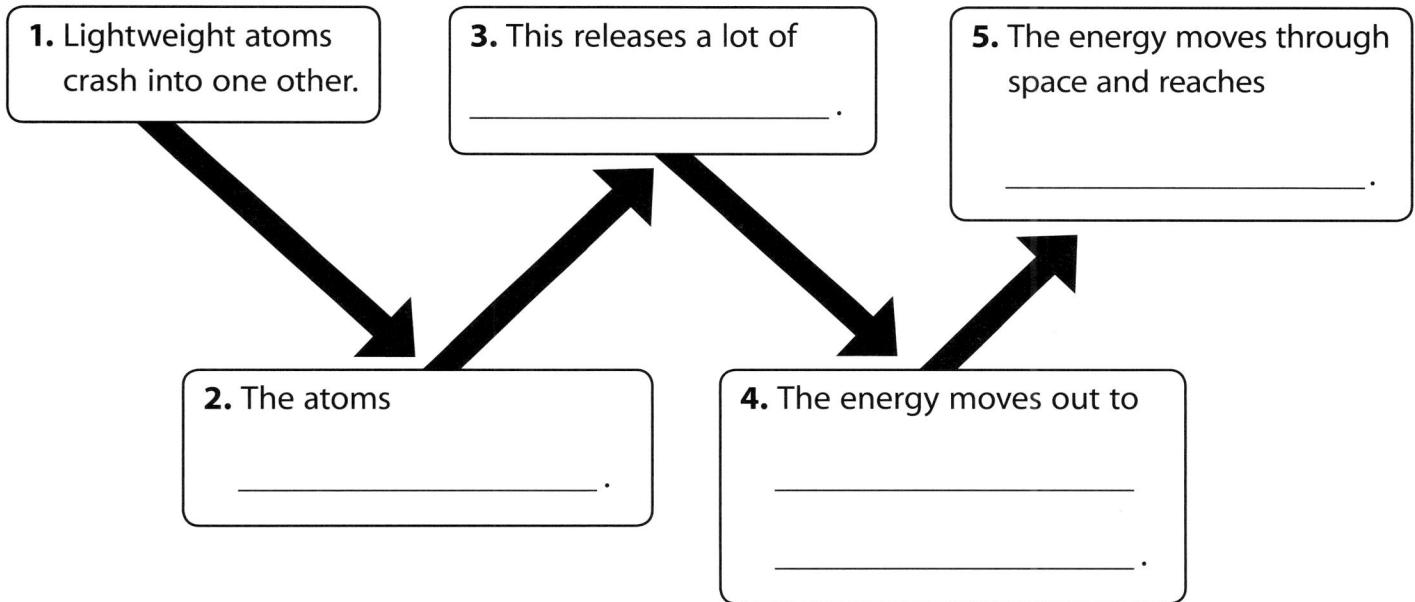

1. Lightweight atoms crash into one other.

3. This releases a lot of _____ .

5. The energy moves through space and reaches _____ .

2. The atoms _____ .

4. The energy moves out to _____ _____ .

STRATEGY FOCUS: ASK QUESTIONS

After reading the chapter, what is one question you still have about stars?

_____

_____

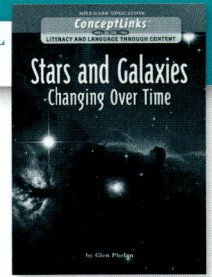

ORANGE LEVEL
pages 4–8

ConceptLinks
LITERACY AND LANGUAGE THROUGH CONTENT
Stars and Galaxies
-Changing Over Time
by Glen Phelan

OBJECTIVES

Students will:

LANGUAGE

SPEAK and **LISTEN** to share ideas about stars.

LITERACY

READ and **ANALYZE** nonfiction texts.

INTERPRET diagrams and charts.

ASK questions to deepen understanding.

CONTENT

UNDERSTAND that all stars shine because fusion releases energy.

UNDERSTAND that stars have different properties, such as temperature and brightness.

MATERIALS

- **Student Book**, Orange Level
- **Learning Master 7**
- **Audio CD**, Track 15

FOR MORE SUPPORT

PAGES 6-7 Point out these phrases: **in fact** in the second paragraph on page 6 and **for example** in the second paragraph on page 7. Explain that they link ideas between sentences or paragraphs. Read aloud each sentence that contains the phrase, and explain them as follows: **In fact** tells the reader that this sentence will add to or explain information. **For example** tells the reader that the sentence will have an example.

Guide Reading

Activity Overview

Orange Level	small group	5-10 minutes

- Use these leveled notes to guide the reading of Chapter 1 in the Orange Level **Student Book**.
- Have the other book-level small groups work on their own Chapter 1 activities. See the detailed grouping plan on page 22.
- When you leave the group, remind students to complete **Learning Master 7** and the Your Turn activities on page 9 of the **Student Book**.
- Return to page 23 of this guide for answers to the Your Turn activities and additional teaching suggestions.

Read Chapter 1: A Look at Stars

Pages 4-5 Read aloud the title and caption. Ask: **What does the sun give us?** (light and heat) Read the definitions of **atoms** and **fusion**. Explain that in this book, important words will be bold and defined on the page when they appear for the first time. Ask: **What is released when fusion occurs?** (energy) **How does this energy move?** (It moves to the surface of the sun and then out into space.) Explain that we feel the effects of fusion as light and heat that reach Earth.

Page 6 Read aloud the definition of **light-year**. Point out that scientists find it easier to work with smaller numbers and larger units of distance than larger numbers and smaller units of distance; for example, 30 kilometers is easier to work with than 30,000 meters. Say: **Suppose it is daytime. The sun suddenly stops shining. How long would it be before our sky goes dark?** (8 minutes)

Page 7 Have students trace the Big Dipper star pattern on the chart. Ask: **Are the stars in the Big Dipper close together in space?** (no) **How do you know?** (The chart shows that they are different distances from Earth.)

BY THE WAY Show students a picture of Ursa Major. Trace the handle of the Big Dipper and show that it is the bear's tail. Ask students to name other constellations they know. (Possible response: Orion)

Page 8 Direct students' attention to the chart. Ask: **What causes a star to be a certain color?** (the surface temperature of the star) Explain that **apparent** means something that appears a certain way but might not be what it seems. Say: **Suppose the stars in this chart are all the same size. Which star has the greatest absolute magnitude?** (the blue star) **What needs to be the same about two stars to compare their absolute magnitudes?** (size and distance from Earth)

Apply the Comprehension Strategy

ASK QUESTIONS Remind students that asking questions as they read helps them learn more. Ask: **What question do you have about a diagram in this chapter?** (Possible response: Why are some stars hotter than others?) Tell students to look for the answers as they read the rest of the book.

Study Guide
Learning Master 7

Stars and Galaxies: Changing Over Time
Chapter 1: A Look at Stars

ORANGE LEVEL
Student Book,
pages 4-8

USE KEY WORDS

Look at the Key Words on page 23 of your book.
Answer these questions about the Key Words in Chapter 1.
Select the best answer.

KEY WORDS
absolute magnitude
apparent magnitude
fusion
light-year
star

1. Which describes how bright a **star** looks?
   - **A.** fusion
   - **C.** absolute magnitude
   - **B.** light-year
   - **D.** apparent magnitude

2. Which describes how bright a **star** really is?
   - **A.** fusion
   - **C.** absolute magnitude
   - **B.** light-year
   - **D.** apparent magnitude

3. Which describes the combining of smaller atoms into one larger atom?
   - **A.** fusion
   - **C.** absolute magnitude
   - **B.** light-year
   - **D.** apparent magnitude

ORGANIZE IDEAS

As you read Chapter 1, complete the chart.

PROPERTIES OF STARS		
Property	**Definition**	**Example**
	a measure of how hot or cold something is	
brightness: apparent magnitude		A small, cooler star looks brighter from Earth than other stars because it is closer.
brightness:	how bright a star really is	

STRATEGY FOCUS: ASK QUESTIONS

What is one question you had that was not answered in this chapter?

_____

_____

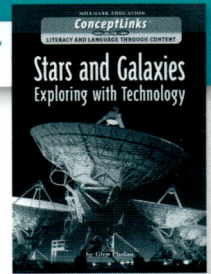

MILMARK EDUCATION
ConceptLinks
LITERACY AND LANGUAGE THROUGH CONTENT
Stars and Galaxies
Exploring with Technology

OBJECTIVES

Students will:

LANGUAGE

SPEAK and **LISTEN** to compare photos of the Crab Nebula.

USE knowledge of Latin word roots to determine meaning.

LITERACY

READ and **ANALYZE** nonfiction texts.

INTERPRET diagrams and charts.

ASK questions to deepen understanding.

CONTENT

UNDERSTAND that stars are huge, glowing balls of hot gases.

UNDERSTAND that scientists use telescopes to collect different kinds of energy, which they study to learn about stars.

MATERIALS

- **Student Book**, Purple Level
- **Learning Master 8**
- **Audio CD**, Track 19

FOR MORE SUPPORT

PAGE 4 Write **astronomer** on the board, and underline *astro-*. Explain that *astro-* comes from a Greek word that means "stars." It also refers to outer space. Ask: **What word with *astro-* means a person who travels in outer space?** (astronaut)

Guide Reading

Activity Overview

Purple Level	small group	5-10 minutes

- Use these leveled notes to guide the reading of Chapter 1 in the Purple Level **Student Book**.
- Have the other book-level small groups work on their own Chapter 1 activities. See the detailed grouping plan on page 22.
- When you leave the group, remind students to complete **Learning Master 8** and the Your Turn activities on page 9 of the **Student Book**.
- Return to page 23 of this guide for answers to the Your Turn activities and additional teaching suggestions.

Read Chapter 1: Collecting Energy from Space

Pages 4-5 Ask: **What kinds of energy do stars give off?** (light and heat) Direct students' attention to the images. Explain that a nebula is a cloud of gas and dust. The star in the nebula shown here gives off energy that is invisible to human eyes and light energy. Ask: **Which image can humans see by looking through a telescope?** (the optical view) Ask: **Why do astronomers use different kinds of telescopes to study the same star?** (They learn different things about the star by studying the different kinds of energy it produces.)

> **SHARE IDEAS** *How are the photos of the Crab Nebula different from each other? Share your ideas.* (Students might mention differences in color, shape, and pattern.)

Page 6 Read aloud the caption, and discuss the design of the Solar Telescope. Ask: **Why should you never look directly at the sun?** (It might damage your eyes.)

> **EXPLORE LANGUAGE** Write these words on the board: **solar panels**, **solar system**, **solarium**. Discuss the meaning of each word.

Page 7 Ask students to explain what is happening in the diagram. (Atoms crash into one another and release a lot of energy.) Write **produce light** and **reflect light** on the board. Have students explain the difference.

> **BY THE WAY** Explain that the energy released in fusion is nuclear energy. Scientists are working on ways to use fusion to supply energy needs. The nuclear energy some power plants use now is produced by fission, in which atoms are split apart.

Page 8 Have students compare the different types of energy that the telescopes collect. Ask: **What is the advantage of having telescopes in space rather than on Earth's surface?** (Earth's air makes stars look blurry and blocks much of the other kinds of energy given off by stars.)

Apply the Comprehension Strategy

ASK QUESTIONS Remind students that they can learn more if they ask questions before, during, and after reading. Ask students to think of questions to ask about the telescopes on page 8.

Stars and Galaxies: Exploring with Technology
Chapter 1: Collecting Energy from Space

PURPLE LEVEL
Student Book,
pages 4-8

USE KEY WORDS

Look at the Key Words on page 23 of your book.
Answer these questions about the Key Words in Chapter 1.

KEY WORDS
fusion
star

1. The combining of atoms is called _____ .

2. A **star** shines because of _____ .

3. What is a **star**?

_____

ORGANIZE IDEAS

As you read Chapter 1, complete the word map.

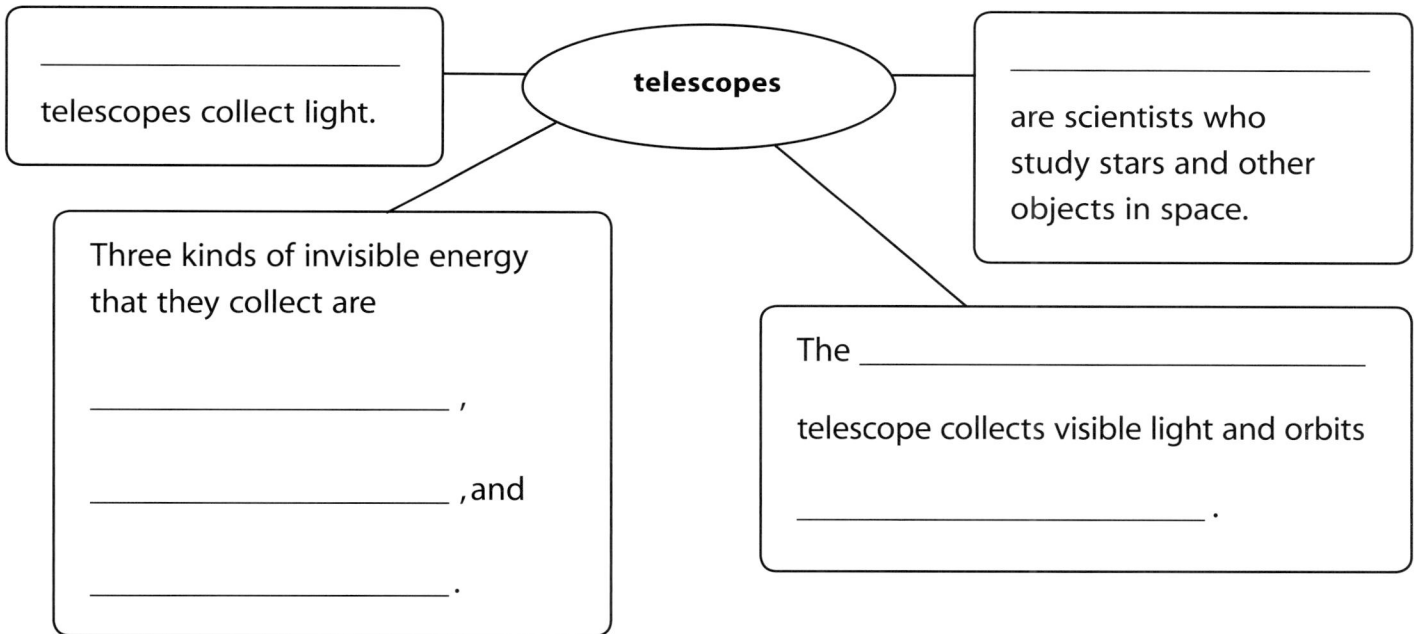

_____

telescopes collect light.

telescopes

_____

are scientists who study stars and other objects in space.

Three kinds of invisible energy that they collect are

_____ ,

_____ ,and

_____ .

The _____

telescope collects visible light and orbits

_____ .

STRATEGY FOCUS: ASK QUESTIONS

What is one question you still have about telescopes?

_____

Where could you look to find the answer to your question?

_____

LESSON ACTIVITIES

Guide Reading
25-40 minutes (TG, pp. 32-41)
· Review Concepts
· Mini-lesson: Summarize
· Read Chapter 2
· Respond and Reflect
· Reinforce and Extend

OBJECTIVES

Students will:

LANGUAGE

USE academic language to ask and answer questions about stars and galaxies.

LITERACY

MAKE CONNECTIONS from text to self and from text to world.

APPLY the comprehension strategy of asking questions.

CONTENT

APPLY the science process skill of summarizing.

See the book-level Guide Reading lessons for additional objectives.

MATERIALS

• **Student Books**, all levels
• **Learning Masters 9-12**
• **Audio CD**, Tracks 8, 12, 16, 20
• **Resource CD-ROM**
• **Image Bank CD-ROM**

Guide Reading

Activity Overview

Review Concepts	whole group	2-3 minutes
Teach the Mini-lesson: Summarize	whole group	5 minutes
Read Chapter 2 Respond and Reflect	small groups small groups	20-30 minutes
Reinforce and Extend	small groups, individuals	20-120 minutes (optional)

Review Concepts

Have students share what they learned about stars and galaxies from reading Chapter 1. Recap the comprehension strategy of asking questions and remind students to ask questions before, during, and after reading.

Teach the Mini-lesson

Science Process Skill: Summarize

ASK STUDENTS to tell what they know about stars and galaxies. List their ideas on the board. If students need ideas, allow them to review the covers of their **Student Books**. Help students **summarize** their information in a chart.

TELL STUDENTS When you summarize, you give a short retelling. You include only the most important ideas. You do not include many details. Using a chart can help you organize the main ideas. Scientists often make charts, diagrams, and models to describe and summarize data. Summarizing makes the information easier to understand.

TELL STUDENTS that they will be summarizing information about stars and galaxies.

Read Chapter 2

Use this grouping plan as a guide for planning the reading time.

BLUE LEVEL	GREEN LEVEL	ORANGE LEVEL	PURPLE LEVEL
Students read with a partner or with the **Audio CD**, Track 8.	**Students read with your guidance.**	Students read the chapter alone or with a partner.	Students read the chapter alone or with a partner.
Students reread with your guidance.	Students reread with a partner or with the **Audio CD**, Track 12.	Students work together to complete the Study Guide, LM 11.	Students work together to complete the Study Guide, LM 12.
Students work together to complete the Study Guide, LM 9.	Students work together to complete the Study Guide, LM 10.	Students begin the Your Turn activities, SB page 15.	**Students reread with your guidance.**
Students begin the Your Turn activities, SB page 15.	Students begin the Your Turn activities, SB page 15.	**Students review with your guidance.**	Students begin the Your Turn activities, SB page 15.

LM = Learning Master   SB = Student Book

- Before students read, distribute the appropriate Study Guides and verify that students understand the tasks.
- As students read, circulate among the small groups to guide the reading and monitor understanding. Use the Guide Reading lessons on pages 34, 36, 38, and 40. Have students complete the book-level Study Guide.
- After students read, have them work in their small groups or with a partner to complete the Your Turn activities on page 15 of the **Student Books**.

Respond and Reflect: Your Turn

Students respond to the reading with hands-on activities that include applying the science process skill, making connections, and applying the comprehension strategy of **asking questions**.

HOME LANGUAGE SUPPORT

Students can deepen their understanding of the Key Words by relating them to home language words. See **E-Masters 11-18** on the **Resource CD-ROM**.

	BLUE LEVEL	GREEN LEVEL	ORANGE LEVEL	PURPLE LEVEL
SUMMARIZE	Students complete sentences that summarize what they learned in the chapter. **Correct response:** The sun is our closest star. Stars look small in the sky because they are so far away. Scientists measure the distance to faraway stars in light-years.	Students make a chart of ways that stars are different from one another, and write or draw an example for each way. **Possible response:** How Stars Are Different: brightness; Example: Larger, hotter stars are brighter than smaller, cooler stars.	Students summarize how stars of average mass and high mass change during their life cycles. **Possible response:** Both kinds of stars form in a nebula. Then they collapse and expand. An average-mass star becomes a red giant and collapses. The outer layers become a planetary nebula. The center becomes a white dwarf. A high-mass star expands into a supergiant, collapses, and explodes into a supernova. It then forms a neutron star or a black hole.	Students make a diagram that shows the steps in the sun's life cycle and write captions that summarize each step. **Correct response:** Students' diagrams should correctly show the steps of an average-mass star's life cycle and include a summary of each step.
MAKE CONNECTIONS	Students explain why the moon looks larger than the stars in the night sky. **Possible response:** The moon looks larger than stars because it is closer to Earth.	Students describe the apparent and absolute magnitudes of lights on the helmets of hikers walking toward them. **Correct response:** The apparent magnitudes will increase as the hikers come closer, but the absolute magnitudes will stay the same.	Students explain why it makes sense for scientists to refer to a star's "life cycle" even though it is not alive. **Possible response:** A star is "born" in a nebula, changes throughout its "life," then "dies" to become a white dwarf, neutron star, or black hole.	Students explain how observing stars is like looking back in time. **Possible response:** It takes years for light to travel from most stars to Earth. What we see happened sometime in the past.
STRATEGY FOCUS: ASK QUESTIONS	Have students ask questions by observing details in the photos, charts, and diagrams, and thinking about what else they would like to know as they read the text.			

Reinforce and Extend: Use Language (optional)

If you have a longer instructional period, students can carry out the Use Language activity on page 21 of the **Student Books**. This activity can be used to initiate the Writer's Workshop. See the teaching notes on page 63 of this guide, as well as the Writer's Workshop on pages 76-81.

ResourceLinks

Students may review the **Image Bank**, Writing Ideas, for images related to the writing prompts.

Lesson Wrap-up Invite students to share their responses to the Your Turn activities with the whole class or in small groups.

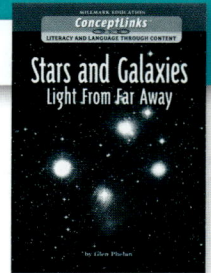

OBJECTIVES

Students will:

LANGUAGE

SPEAK and **LISTEN** to share ideas about how distances to stars are measured.

LITERACY

READ and **ANALYZE** nonfiction texts.

INTERPRET diagrams and charts.

ASK questions to deepen understanding.

CONTENT

UNDERSTAND that stars are huge glowing balls of hot gases and that the distances of all stars except the sun are measured in light-years.

UNDERSTAND that some objects in space reflect light from stars.

MATERIALS

- **Student Book**, Blue Level
- **Learning Master 9**
- **Audio CD**, Track 8

FOR MORE SUPPORT

PAGE 10 Point out the word **closest** in the first sentence. Arrange three objects at varying distances from you, and use the words **close**, **closer**, and **closest** to describe them. Show students how to make the comparative and superlative forms of close by adding -er or -est.

Guide Reading

Activity Overview

Blue Level	small group	5-10 minutes

- Use these leveled notes to guide the reading of Chapter 2 in the Blue Level **Student Book**.
- Have the other book-level small groups work on their own Chapter 2 activities. See the detailed grouping plan on page 32.
- When you leave the group, remind students to complete **Learning Master 9** and the Your Turn activities on page 15 of the **Student Book**.
- Return to page 33 of this guide for answers to the Your Turn activities and additional teaching suggestions.

Read Chapter 2: Distances to Stars

Pages 10-11 Read aloud the title and labels. Discuss the photo on page 10. Ask: **What do you already know about the sun?** (Possible response: It gives off light and heat.) Ask a volunteer to read the Key Ideas. Have students note that the sun is a star. Ask: **What are stars made of?** (hot gases) Direct students' attention to the diagram on page 11. Explain that the distances between objects are not to scale, but the sizes of the objects are to scale. Have students compare the sizes. Ask: **How many Earths can fit into the sun?** (one million) **How does the size of the sun compare to the size of other stars?** (It is bigger than some stars and smaller than others.)

Page 12 Read aloud the caption. Ask: **What is the brightest star in the night sky?** (Sirius) Tell students that Sirius is about twice the size of the sun. Ask: **Why does Sirius look smaller than the sun?** (It is farther away.) Point out the word **average** in the first sentence of the second paragraph. Explain that average means "in the middle." As Earth orbits the sun, its distance to the sun changes. Earth is closest to the sun in January at 147.1 million km (91.3 million miles) and farthest from the sun in June at about 152.1 million km (94.5 million miles). The average of these two numbers is 149.6 million km (92.9 million miles). The numbers are rounded up to the nearest whole number in the book.

Page 13 Ask: **What is the distance light travels in one year called?** (a light-year) **Why do scientists use light-years to measure distances in space?** (Stars are too far away to use kilometers or miles easily.) Have students study the chart. Point out that four light-years means that it takes light four years to reach Earth. Have students find Betelgeuse in the photo. Ask: **How far away is Betelgeuse?** (about 500 light-years) **Which star in the chart is closest to Earth?** (Proxima Centauri). **Which is farthest?** (Deneb)

Page 14 Read aloud the caption and the Key Idea. Ask students to compare the brightness of the moon with that of the sun on page 10. Ask: **Why does the sun look bright?** (It makes its own light.) **Why does the moon look bright?** (It reflects light from the sun.)

Study Guide
Learning Master 9

Stars and Galaxies: Light from Far Away
Chapter 2: Distances to Stars

BLUE LEVEL
Student Book, pages 10-14

USE KEY WORDS

Look at the Key Words on page 23 of your book.
Answer these questions about Key Words in Chapter 2.

KEY WORDS

constellation
light-year
star

1. The sun is the closest _____ to Earth.

2. A group of **stars** makes a pattern called a _____ .

3. A _____ is the distance light travels in one year.

ORGANIZE IDEAS

As you read Chapter 2, fill in the concept map with facts about stars.

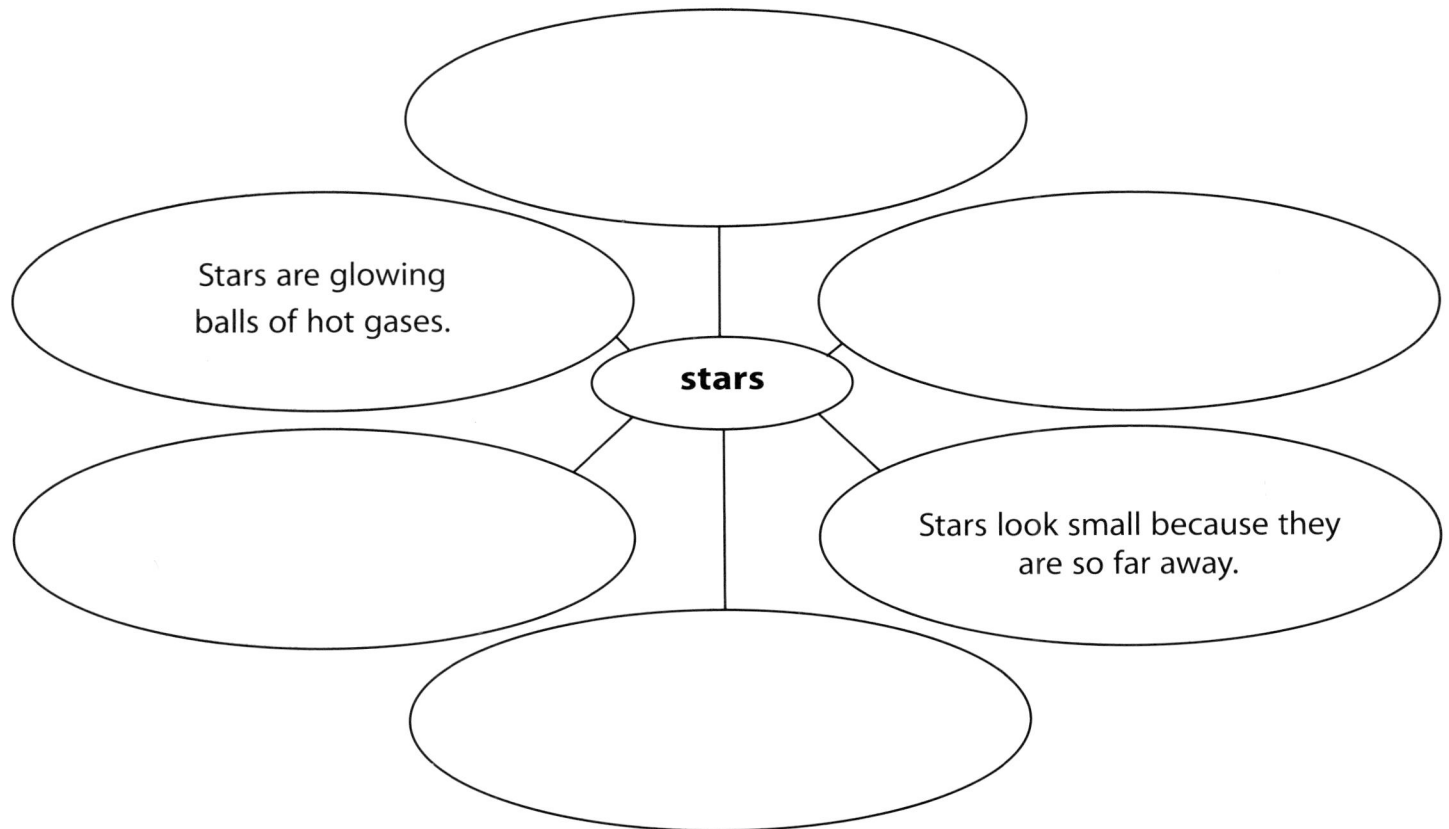

Stars are glowing balls of hot gases.

stars

Stars look small because they are so far away.

STRATEGY FOCUS: ASK QUESTIONS

What is one question you were able to answer while reading this chapter? What is the answer?

_____

_____

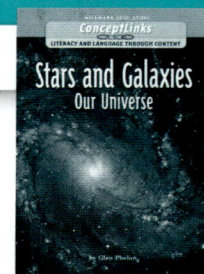

OBJECTIVES

Students will:

SPEAK and **LISTEN** to explain how hot the sun is.

USE descriptive words to tell about the brightness of stars.

READ and **ANALYZE** nonfiction texts.

INTERPRET charts and diagrams.

ASK questions to deepen understanding.

UNDERSTAND that the color of a star is related to its temperature.

UNDERSTAND that stars vary greatly in brightness and size.

MATERIALS

• **Student Book**, Green Level
• **Learning Master 10**
• **Audio CD**, Track 12

FOR MORE SUPPORT

PAGE 10 Read aloud the last paragraph. Point out the words **first**, **as**, **then**, and **finally**. Discuss how these words help explain the order of steps. The word **as** means something is happening at the same time as something else is happening. Have students practice using these words by telling an event in sequence.

Guide Reading

Activity Overview

Green Level	small group	5-10 minutes

• Use these leveled notes to guide the reading of Chapter 2 in the Green Level **Student Book**.
• Have the other book-level small groups work on their own Chapter 2 activities. See the detailed grouping plan on page 32.
• When you leave the group, remind students to complete **Learning Master 10** and the Your Turn activities on page 15 of the **Student Book**.
• Return to page 33 of this guide for answers to the Your Turn activities and additional teaching suggestions.

Read Chapter 2: Different Kinds of Stars

Pages 10-11 Have students read the title and the Key Idea. Ask: **Why are stars different colors**? (Some stars are hotter than others.) **Why are some stars hotter than others?** (Hotter stars have more atoms combining in them. They release more energy.) Direct students' attention to the chart. Ask: **Why is the sun described as medium-hot?** (The sun is yellow. This color means that some stars are hotter and some stars are colder than the sun.)

> **SHARE IDEAS** *How hot is the sun?* **Explain** *how you know.* (The chart shows me that the temperature of the sun is 5,200-6,000°C.)

Pages 12-13 Write **apparent magnitude** and **absolute magnitude** on the board. Have students tell details about each term. Write the details under the appropriate term. Discuss the photo on page 12. Say: **Let's assume all the lights have the same kind of bulb. Which light has the greatest apparent magnitude?** (the brightest light; the one in front) **Does this light have the greatest absolute magnitude? Why or why not?** (No; all

the lights have the same absolute magnitude. The brightest light looks brighter because it is closest.) Have students study the image on page 13. Say: **Adhara has a higher absolute magnitude than Sirius. Why does Sirius look brighter?** (Sirius is closer to Earth.)

> **EXPLORE LANGUAGE** Read the definitions of **apparent** and **absolute**. Give examples, such as the passage of time—some regular school days seem to go faster than others (apparent), but their lengths are the same (absolute).

Page 14 Read aloud the Key Idea. Have students read the labels in the diagram. Ask: **Which star is the biggest?** (the red supergiant) Ask students to think about the size of Earth compared to the size of the stars shown in the diagram. Ask: **How might we show Earth on this diagram?** (Possible response: as a dot)

> **SHARE IDEAS** *What are some common objects you could use to* **compare** *the sizes of stars?* (Possible response: different kinds of balls, such as a ping pong ball, baseball, and basketball)

Stars and Galaxies: Our Universe
Chapter 2: Different Kinds of Stars

GREEN LEVEL
Student Book,
pages 10-14

USE KEY WORDS

Look at the Key Words on page 23 of your book.
Answer these questions about the Key Words in Chapter 2.
Select the best answer.

KEY WORDS
absolute magnitude
apparent magnitude

1. What does **absolute magnitude** describe?

   A. how bright a star really is   **C.** how bright a star looks

   B. how close a star really is   **D.** how close a star looks

2. The closer a star is to Earth, the _____ (higher/lower) the star's
apparent magnitude.

ORGANIZE IDEAS

As you read Chapter 2, complete the chart. First, read the cause. Then, write the effect.

Cause	Effect
1. Hotter stars have more atoms combining in them.	**1.**
2. Stars have different temperatures.	**2.**
3. A star with high absolute magnitude is far away from Earth.	**3.** The star looks _____ from Earth and has a _____ apparent magnitude.
4. A star with low absolute magnitude is very close to Earth.	**4.** The star looks _____ from Earth and has a _____ apparent magnitude.

STRATEGY FOCUS: ASK QUESTIONS

What new question do you have about the chapter?

_____

_____

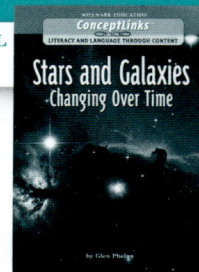

ConceptLinks
LITERACY AND LANGUAGE THROUGH CONTENT

Stars and Galaxies
Changing Over Time

by Glen Phelps

OBJECTIVES

Students will:

LANGUAGE

USE knowledge of Latin words to determine meaning.

LITERACY

READ and **ANALYZE** nonfiction texts.

INTERPRET diagrams.

ASK questions to deepen understanding.

CONTENT

UNDERSTAND that stars form as gravity pulls gas and dust together.

UNDERSTAND that a star changes greatly as it goes through its life cycle.

MATERIALS

- **Student Book**, Orange Level
- **Learning Master 11**
- **Audio CD**, Track 16

FOR MORE SUPPORT

PAGE 11 Read aloud the second sentence. Explain that **average** can mean "in the middle." A star with average mass has a mass about halfway between a small mass and large mass. Help students think of other ways they use this word, such as average height and the average of a group of numbers.

Guide Reading

Activity Overview

Orange Level	small group	5-10 minutes

- Use these leveled notes to guide the reading of Chapter 2 in the Orange Level **Student Book**.
- Have the other book-level small groups work on their own Chapter 2 activities. See the detailed grouping plan on page 32.
- When you leave the group, remind students to complete **Learning Master 11** and the Your Turn activities on page 15 of the **Student Book**.
- Return to page 33 of this guide for answers to the Your Turn activities and additional teaching suggestions.

Read Chapter 2: The Life Cycle of a Star

Page 10 Read aloud the title, caption, and label. Have students describe what they see in the photo and read the definitions. Write the word **nebula** on the board. Hold up a sheet of paper. Tell the students that the paper represents gases in space. Place your hand on the paper. Make a fist as you pull the paper into a ball. Explain that in space the pulling force of gravity causes a star to form from the gases, similar to the way that your fist pulled the paper into a ball.

Page 11 Explain that we talk about stars as having "lives" and "life cycles." Ask students to name parts of a person's life cycle. (Possible response: being born, growing up, growing old, and dying) Point out that stars change over time in the same way, but they are not alive. Ask: **What does the length of a star's life depend on?** (the mass of the star) **The brightest star in the night sky has about twice the mass of the sun. Which star will live longer?** (the sun) **How long is the sun's life?** (ten billion years)

EXPLORE LANGUAGE Write the word **nebulizer** on the board. Ask students if they have ever used nose spray or perfume. Then tell them that a nebulizer is a device that sprays medicine in the form of a mist into the nose.

Pages 12-13 Have students follow along as volunteers read the captions from beginning to end for the star of average mass, and then for the star of high mass. Have students act out the steps as they say aloud each step. For example, students might spread their arms wide for **expand** and hug their arms around their bodies for **collapse**. Ask: **Which steps are alike for both kinds of stars?** (the first two steps) Make sure students understand that the changes shown in this diagram do not represent the entire life cycle.

Page 14 Read aloud the Key Idea and caption. Ask: **What happens to the remaining material after a supernova?** (It might become a neutron star or a black hole.) **What kind of star becomes a black hole?** (a star with at least three times the mass of the sun) **What would happen if a spacecraft were pulled into a black hole?** (It would not be able to escape.)

NAME ______________________________

Study Guide
Learning Master 11

Stars and Galaxies: Changing Over Time
Chapter 2: The Life Cycle of a Star

ORANGE LEVEL
Student Book, pages 10-14

USE KEY WORDS

KEY WORDS
black hole
gravity
mass
nebula
red giant
supernova

Look at the Key Words on page 23 of your book.
Answer these questions about the Key Words in Chapter 2.
Select the best answer.

1. What is the name for a huge cloud of gas and dust in space?
A. black hole
B. nebula
C. supernova
D. red giant

2. What does a dying supergiant star become?
A. a black hole
B. a nebula
C. a supernova
D. a red giant

3. Which phrase describes a **black hole**?
A. strong gravity
B. an explosion
C. a cloud of gas and dust
D. a large, cool star

ORGANIZE IDEAS

As you read Chapter 2, complete the chart to compare the end of the life cycle of a star of average mass with a star of high mass.

What happens	Star of average mass	Star of high mass
forms in a ____________	nebula	
____________ the first time		collapses
expands to become a ____________		
____________ the second time		
expands to become a ____________		
center or remainder collapses to become a ____________		neutron star or

STRATEGY FOCUS: ASK QUESTIONS

What is one question you were able to answer while reading this chapter? What is the answer?

____

____

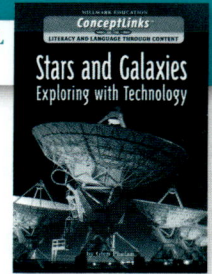

EXPLORE CHAPTER 2 (cont'd)

PURPLE LEVEL
pages 10-14

CONCEPTLINKS
NILLMAN EDUCATION
LITERACY AND LANGUAGE THROUGH CONTENT
Stars and Galaxies
Exploring with Technology

OBJECTIVES

Students will:

LANGUAGE

SPEAK and **LISTEN** to share ideas about exploring stars.

LITERACY

READ and **ANALYZE** nonfiction texts.

INTERPRET charts.

ASK questions to deepen understanding.

CONTENT

UNDERSTAND that distances to stars are measured in light-years.

UNDERSTAND that the properties of a star change as it goes through its life cycle.

MATERIALS

- **Student Book**, Purple Level
- **Learning Master 12**
- **Audio CD**, Track 20

FOR MORE SUPPORT

PAGE 11 Point out the word **brightness** in the second paragraph. Explain that **bright** is an adjective. Adding the suffix -ness turns the adjective into a noun. Write these words on the board: **careless**, **fair**, **gentle**. Have students turn them into nouns by adding -ness. Discuss the meanings of the new words.

Guide Reading

Activity Overview

Purple Level	small group	5-10 minutes

- Use these leveled notes to guide the reading of Chapter 2 in the Purple Level **Student Book**.
- Have the other book-level small groups work on their own Chapter 2 activities. See the detailed grouping plan on page 32.
- When you leave the group, remind students to complete **Learning Master 12** and the Your Turn activities on page 15 of the **Student Book**.
- Return to page 33 of this guide for answers to the Your Turn activities and additional teaching suggestions.

Read Chapter 2: Exploring Stars

Page 10 Ask: **What do you notice in the photo?** (Possible response: Some stars are brighter than others. Some stars are different colors.) Display an image that shows how the stars in Orion are connected to make a pattern in the shape of a hunter. Ask: **What is this pattern called?** (a constellation) Make sure students understand that a light-year is a measure of distance, not time.

Page 11 Use the chart to point out the relationship between star color and temperature. Ask: **What do you notice about the color of a star and different surface temperatures?** (The colors are different for different temperatures.) Ask students to describe the change in color as the temperature increases. (Possible response: The colors get farther from red and closer to blue.) Dim the lights in the classroom. Have volunteers shine lamps or flashlights with bulbs of equal brightness at a wall while standing at different distances from the wall. Have students describe the lights in terms of apparent magnitude and absolute magnitude. Ask: **How**

could we make the absolute magnitude of one light different from another light? (Use a brighter or a dimmer bulb.) Read the Key Ideas. Ask: **What are three properties of stars?** (color, surface temperature, size)

Pages 12-13 Explain that stars change over time. Have students look at the photo on page 12 and read the caption. Have students describe the sequence of events that result in the birth of a star. Then have them look at the diagram on page 13. Have volunteers read the life cycles of stars with different masses. At the appropriate times, pause the reader and help students locate the stage in the diagram. Ask: **What properties of a star change as it goes through its life cycle?** (shape, temperature, color, size, and brightness)

Page 14 Have students read the definitions and look at the photo. Ask: **If scientists can't see a black hole, how do they find it?** (They observe matter or light being pulled into a dark area of space.) **What kind of energy do black holes give off?** (X-rays) **What kind of energy do pulsars give off?** (radio energy) **What kind of instrument might astronomers use to detect a pulsar?** (a radio telescope)

Stars and Galaxies: Exploring with Technology
Chapter 2: Exploring Stars

USE KEY WORDS

Look at the Key Words on page 23 of your book.

Answer these questions about the Key Words in Chapter 2.

Select the best answer.

KEY WORDS
black hole
gravity
light-year
mass
neutron star
pulsar
red giant
supernova

1. A large, cool star that has expanded from a star of average mass is called a _____ .

   A. red giant   **C.** light-year

   B. supernova   **D.** pulsar

2. Stars form when _____ pulls gas and dust together.

   A. a black hole   **C.** gravity

   B. a supernova   **D.** mass

3. A rapidly spinning **neutron star** that gives off bursts of radio waves is called a _____ .

   A. supernova   **C.** black hole

   B. pulsar   **D.** red giant

ORGANIZE IDEAS

As you read Chapter 2, write the correct terms in the boxes to show the differences between the life cycles of a star with an average mass and a star with a high mass.

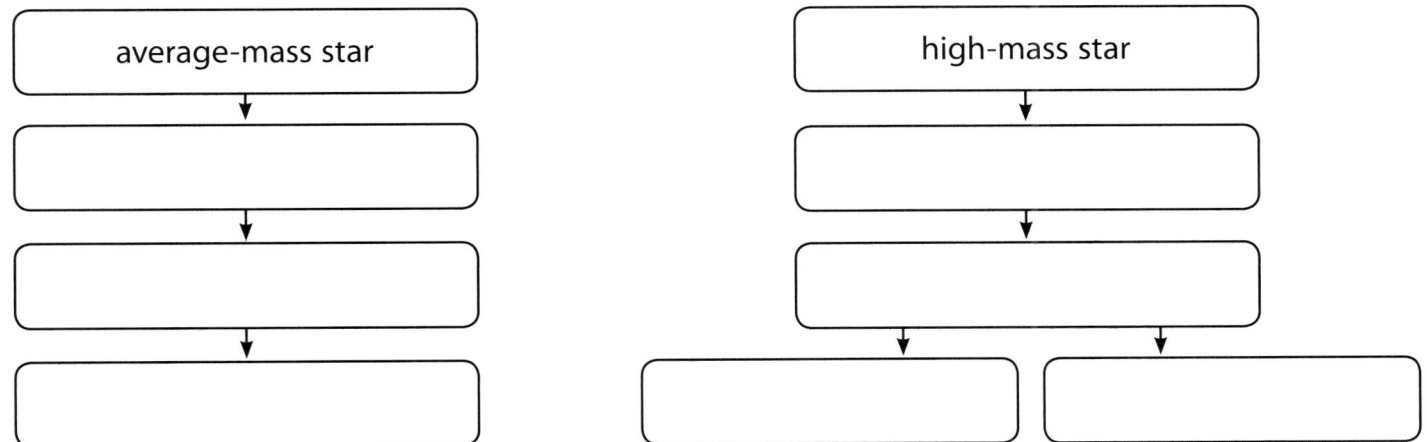

average-mass star

↓

↓

↓

high-mass star

↓

↓

↓

STRATEGY FOCUS: ASK QUESTIONS

What is one question you still have about black holes?

LESSON ACTIVITIES

Guide Reading
25-40 minutes (TG, pp. 42-51)
- Review Concepts
- Mini-lesson: Infer
- Read Chapter 3
- Respond and Reflect
- Reinforce and Extend

OBJECTIVES

Students will:

LANGUAGE

USE academic language to ask and answer questions about stars and galaxies.

EXPAND knowledge of content-area words, including multiple-meaning words.

LITERACY

MAKE CONNECTIONS from text to self and from text to world.

APPLY the comprehension strategy of asking questions.

CONTENT

APPLY the science process skill of inferring.

See the book-level Guide Reading lessons for additional objectives.

MATERIALS

- **Student Books**, all levels
- **Learning Masters 13-16**
- **Audio CD**, Tracks 9, 13, 17, 21
- **Image Bank CD-ROM**
- **Resource CD-ROM**

ALL LEVELS

Guide Reading

Activity Overview

Review Concepts	whole group	2-3 minutes
Teach the Mini-lesson: Infer	whole group	5 minutes
Read Chapter 3 Respond and Reflect	small groups small groups	20-30 minutes
Reinforce and Extend	small groups, individuals	5-15 minutes (optional)

Review Concepts

Have students share what they learned about stars and galaxies from reading Chapter 2. Have them restate the Key Ideas in their own words.

Teach the Mini-lesson

Science Process Skill: Infer

DISPLAY a photo from a newspaper or a magazine that shows people with expressive faces. Examples include people smiling or laughing at a picnic, people cheering at a sporting event, or people with expressions of shock or wonder. Have students explain what they think is happening to cause those expressions.

TELL STUDENTS When you infer, you use what you know plus clues from what you read, see, or hear to understand what you are reading, seeing, or hearing. You made inferences about the photo. Scientists make inferences, too, based on their observations. Then they test whether their inferences are correct.

TELL STUDENTS that they will be making inferences about stars and galaxies.

Read Chapter 3

Use this grouping plan as a guide for planning reading time.

BLUE LEVEL	GREEN LEVEL	ORANGE LEVEL	PURPLE LEVEL
Students read with a partner or with the **Audio CD**, Track 9.	Students read the chapter alone or with a partner.	Students read the chapter alone or with a partner.	**Students read with your guidance.**
Students work together to complete the Study Guide, LM 13.	Students work together to complete the Study Guide, LM 14.	**Students reread with your guidance.**	Students reread the chapter alone or with a partner.
Students reread with your guidance.	Students begin the Your Turn activities, SB page 19.	Students work together to complete the Study Guide, LM 15.	Students work together to complete the Study Guide, LM 16.
Students begin the Your Turn activities, SB page 19.	**Students review with your guidance.**	Students begin the Your Turn activities, SB page 19.	Students begin the Your Turn activities, SB page 19.

LM = Learning Master   SB = Student Book

- Before students read, distribute the appropriate Study Guides and verify that students understand the tasks.
- As students read, circulate among the small groups to guide the reading and monitor understanding. Use the Guide Reading lessons on pages 44, 46, 48, and 50. Have students complete the book-level Study Guide.
- After students read, have them work in their small groups or with a partner to complete the Your Turn activities on page 19 of the **Student Books**.

HOME LANGUAGE SUPPORT

Invite students to discuss the Key Ideas from the chapter in their home language. For home language resources, see the **Resource CD-ROM**.

Respond and Reflect: Your Turn

Students respond to the reading with hands-on activities that include applying the science process skill, making connections, and expanding their vocabulary.

	BLUE LEVEL	GREEN LEVEL	ORANGE LEVEL	PURPLE LEVEL
INFER	Students look at a photo to infer the distances between galaxies. **Possible response:** These galaxies must be many light-years apart.	Students look at a photo and diagram of the Milky Way and infer why the stars in the Milky Way look so close together. **Possible response:** The stars of the Milky Way look close together because you are looking at the stars from Earth.	Students make inferences about what a supernova and a white dwarf each was in the past. **Possible response:** I can infer that the supernova was a supergiant. I can infer that the white dwarf was a red giant.	Students use a diagram to infer the position from which they are looking at the Milky Way. **Possible response:** I am looking along the A arrow. Earth is located near the sun. From this position, the many stars in the center of our galaxy look like a cloudy patch of light.
MAKE CONNECTIONS	Students explain why they think our galaxy is called the Milky Way. **Possible response:** The Milky Way Galaxy looks white in the night sky, like a band of spilled milk.	Students explain how the names of galaxies are related to their shapes. **Correct response:** Spiral galaxies have arms that spiral outward. Elliptical galaxies have the shape of a slightly flat circle, or an ellipse. Irregular galaxies have no definite shape.	Students name things that are classified into groups and describe the traits used to classify them. **Possible response:** Students are classified by grade according to their ages.	Students explain why rising raisin bread dough is similar to the expanding universe. **Correct response:** As the dough rises, it expands and the raisins spread out. As the universe expands, galaxies spread out.
EXPAND VOCABULARY	Students find the meanings of number words, write each word as a numeral and tell their order. There are one thousand billions in one trillion. (See the **Image Bank CD-ROM**.)	Students use a thesaurus to find synonyms for words. (See the **Image Bank CD-ROM**.)	Students write and draw pictures to explain different meanings of **spiral**. (See the **Image Bank CD-ROM**.)	Students find out what the words **pulsar**, **pulse**, and **pulsates** mean. Then they write and draw pictures to show how the words are connected. (See the **Image Bank CD-ROM**.)

Reinforce and Extend: Science Around You (optional)

If you have a longer instructional period, students can carry out the Science Around You activity on page 22 of the **Student Books**. See the teaching notes on page 63 of this guide.

ResourceLinks

See the **Image Bank** for downloadable versions of photos related to the activities.

Lesson Wrap-up Invite students to share their responses to the Your Turn activities with the whole class or in small groups.

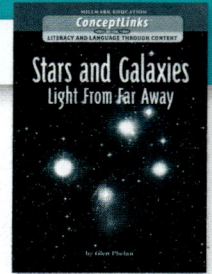

ConceptLinks
Stars and Galaxies
Light From Far Away

OBJECTIVES

Students will:

LANGUAGE

SPEAK and **LISTEN** to share ideas about galaxies.

LITERACY

READ and **ANALYZE** nonfiction texts.

USE visuals to aid comprehension.

ASK questions to deepen understanding.

CONTENT

UNDERSTAND that a galaxy is a huge group of stars held together by gravity.

UNDERSTAND that our sun and Earth are in the Milky Way Galaxy.

MATERIALS

- **Student Book**, Green Level
- **Learning Master 14**
- **Audio CD**, Track 13

FOR MORE SUPPORT

PAGE 17 Read aloud the second sentence. Write it on the board and underline **that pulls things toward each other**. Explain that the part of a sentence that begins with **that** is describing the word before it, in this case, **force**. Have students use this pattern to make up their own sentences.

Guide Reading

Activity Overview

Blue Level	small group	5-10 minutes

- Use these leveled notes to guide the reading of Chapter 3 in the Blue Level **Student Book**.
- Have the other book-level small groups work on their own Chapter 3 activities. See the detailed grouping plan on page 42.
- When you leave the group, remind students to complete **Learning Master 13** and the Your Turn activities on page 19 of the **Student Book**.
- Return to page 43 of this guide for answers to the Your Turn activities and additional teaching suggestions.

Read Chapter 3: Groups of Stars

Pages 16-17 Read the title, caption, and Key Idea. Have students describe what they see in the photos. Discuss the colors and the differences in star brightness. Have students drop a pencil on their desk. Ask: **Why did the pencil fall?** (Gravity pulled it down.) Explain that gravity is a force that pulls things toward each other. The dropped pencil pulls on Earth. But because Earth is so much bigger than the pencil, the pencil's force doesn't affect Earth as much as Earth affects the pencil. Point out that all objects pull toward each other because of gravity. The force between small objects is very weak, so we do not notice it. Ask: **How is gravity important to a galaxy?** (It holds the parts of a galaxy together.)

Page 18 Read the caption and Key Idea. Ask: **What is our galaxy called?** (the Milky Way Galaxy) Draw students' attention to the photo. Have them describe what they see. Explain that the sun is just one of billions of stars that are part of the Milky Way Galaxy. Have students recall how distances to stars are measured and what scientists use to see faraway stars. Ask: **How long do you think it would take to travel from one end of the Milky Way Galaxy to the other in light-years?** (Students' responses should indicate that the number would be very large.) **How do you know?** (The Milky Way Galaxy contains billions of stars, and stars are light-years from one another.) **How do scientists see stars in other galaxies?** (They use powerful telescopes that can see deep into space.)

Study Guide
Learning Master 13

Stars and Galaxies: Light from Far Away
Chapter 3: Groups of Stars

BLUE LEVEL
Student Book,
pages 16-18

USE KEY WORDS

Look at the Key Words on page 23 of your book.
Answer these questions about the Key Words in Chapter 3.

KEY WORDS
galaxy
gravity
Milky Way Galaxy
stars

1. A **galaxy** is a group of _____ , gas, and dust.

2. **Stars** you can see with just your eyes are in the _____ .

3. How does **gravity** help form a **galaxy**?

_____

ORGANIZE IDEAS

As you read Chapter 3, answer each question in the concept map.

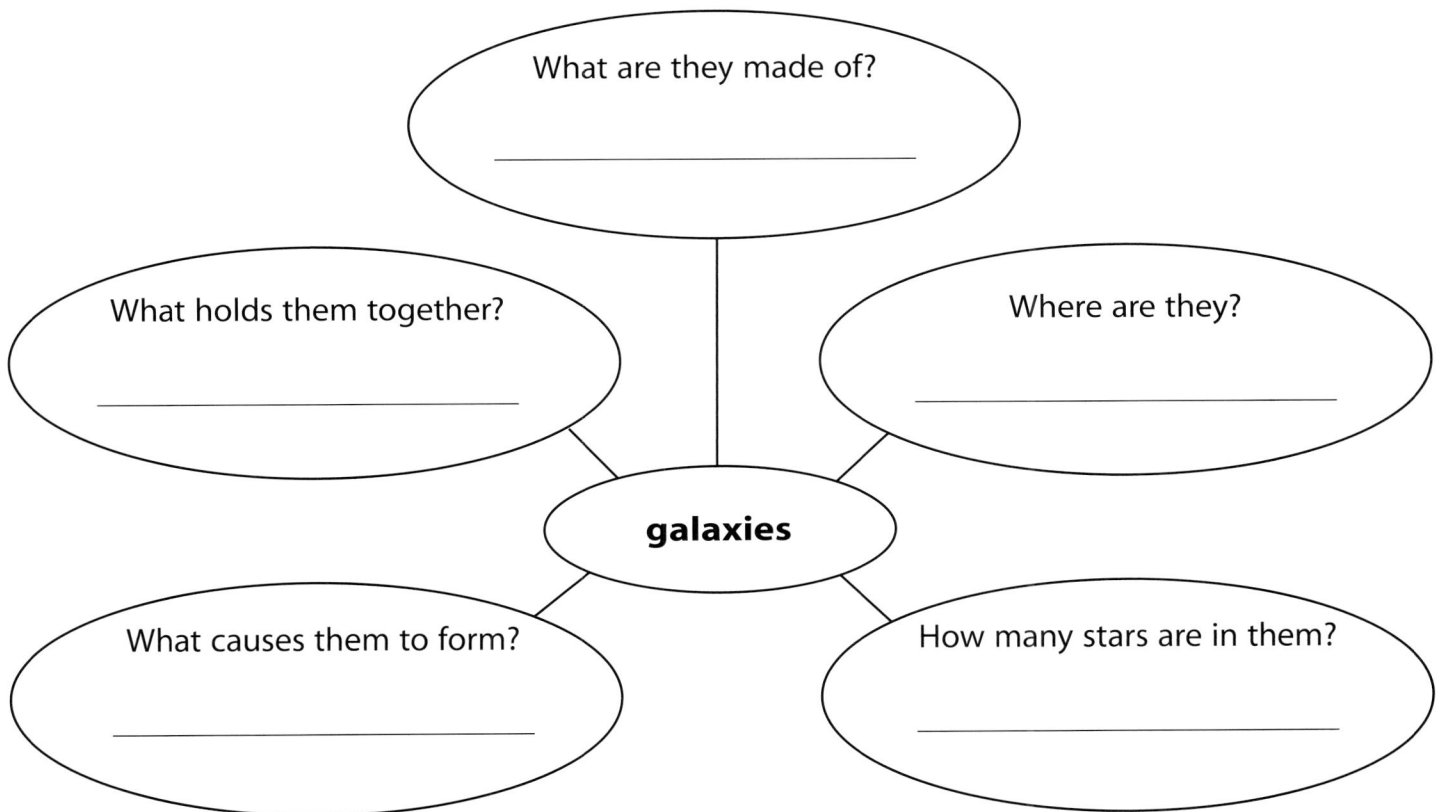

What are they made of?
_____

What holds them together?
_____

Where are they?
_____

galaxies

What causes them to form?
_____

How many stars are in them?
_____

STRATEGY FOCUS: ASK QUESTIONS

What is one question you still have about galaxies?

_____

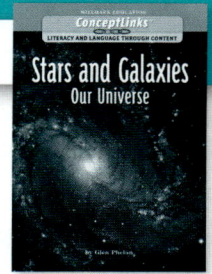

Stars and Galaxies
Our Universe

OBJECTIVES

Students will:

LANGUAGE

SPEAK and **LISTEN** to share ideas about what galaxies are and how they are classified.

LITERACY

READ and **ANALYZE** nonfiction texts.

USE visuals to aid comprehension.

ASK questions to deepen understanding.

CONTENT

UNDERSTAND that a galaxy is a huge group of stars held together by gravity.

UNDERSTAND that galaxies are classified by their shapes.

MATERIALS

- **Student Book**, Green Level
- **Learning Master 14**
- **Audio CD**, Track 13

FOR MORE SUPPORT

PAGE 16 Point out the image of the elliptical galaxy and read the label. Draw the shape on the board. Explain that **elliptical** describes the shape of an oval or a slightly flattened circle. Help students to see that the image of the galaxy is not perfectly round.

Guide Reading

Activity Overview

Green Level	small group	5-10 minutes

- Use these leveled notes to guide the reading of Chapter 3 in the Green Level **Student Book**.
- Have the other book-level small groups work on their own Chapter 3 activities. See the detailed grouping plan on page 42.
- When you leave the group, remind students to complete **Learning Master 14** and the Your Turn activities on page 19 of the **Student Book**.
- Return to page 43 of this guide for answers to the Your Turn activities and additional teaching suggestions.

Read Chapter 3: A Universe of Galaxies

Page 16 Point out the images, and have students compare and contrast their appearances. Explain that scientists classify galaxies by their shape. Draw several regular shapes, such as a circle, a square, and a triangle, on the board. Explain that these are regular shapes because the sides and angles are the same, or one side matches the other. Point out the word **irregular**, and explain that the prefix *ir-* means "not." Draw an irregular shape, and explain that irregular means "not regular." Have students explain the differences between the two kinds of shapes. Ask: **What is true about all of these galaxies?** (They are groups of stars, gas, and dust held together by gravity.)

Page 17 Read the caption. Have students explain what they know about telescopes. If possible, display a telescope in class. Remind students that the light collected by telescopes today came from stars many light-years away and is just now reaching Earth. Ask: **Using telescopes, what have scientists learned about how galaxies are grouped?**

(They have learned that galaxies are grouped in clusters.) Reinforce the meaning of **cluster** by telling students they are a cluster of students.

BY THE WAY Scientists estimate that in about 3 billion years, the Milky Way Galaxy and the Andromeda Galaxy, another spiral galaxy, may collide. The two galaxies are approximately 2 million light-years apart.

Page 18 Read the Key Ideas and the caption. Have a student read the definitions. Ask: **What is the name of the galaxy that includes all the stars that we can see without a telescope?** (the Milky Way Galaxy) Draw students' attention to the photo. Have students describe what they see. Ask: **Why can we see the Milky Way Galaxy without a telescope?** (Earth is part of the Milky Way Galaxy.)

Study Guide
Learning Master 14

Stars and Galaxies: Our Universe
Chapter 3: A Universe of Galaxies

GREEN LEVEL
Student Book,
pages 16-18

USE KEY WORDS

Look at the Key Words on page 23 of your book.
Answer these questions about the Key Words in Chapter 3.

KEY WORDS

galaxy
gravity
Milky Way Galaxy
telescopes
universe

1. A _____ forms when **gravity** pulls stars together.

2. The **Milky Way Galaxy** is part of the _____ .

3. What do scientists use **telescopes** to study?

_____

ORGANIZE IDEAS

As you read Chapter 3, complete the concept map. In each box, write a science word that is connected to galaxies. It does not have to be a Key Word. Then write a definition for the word.

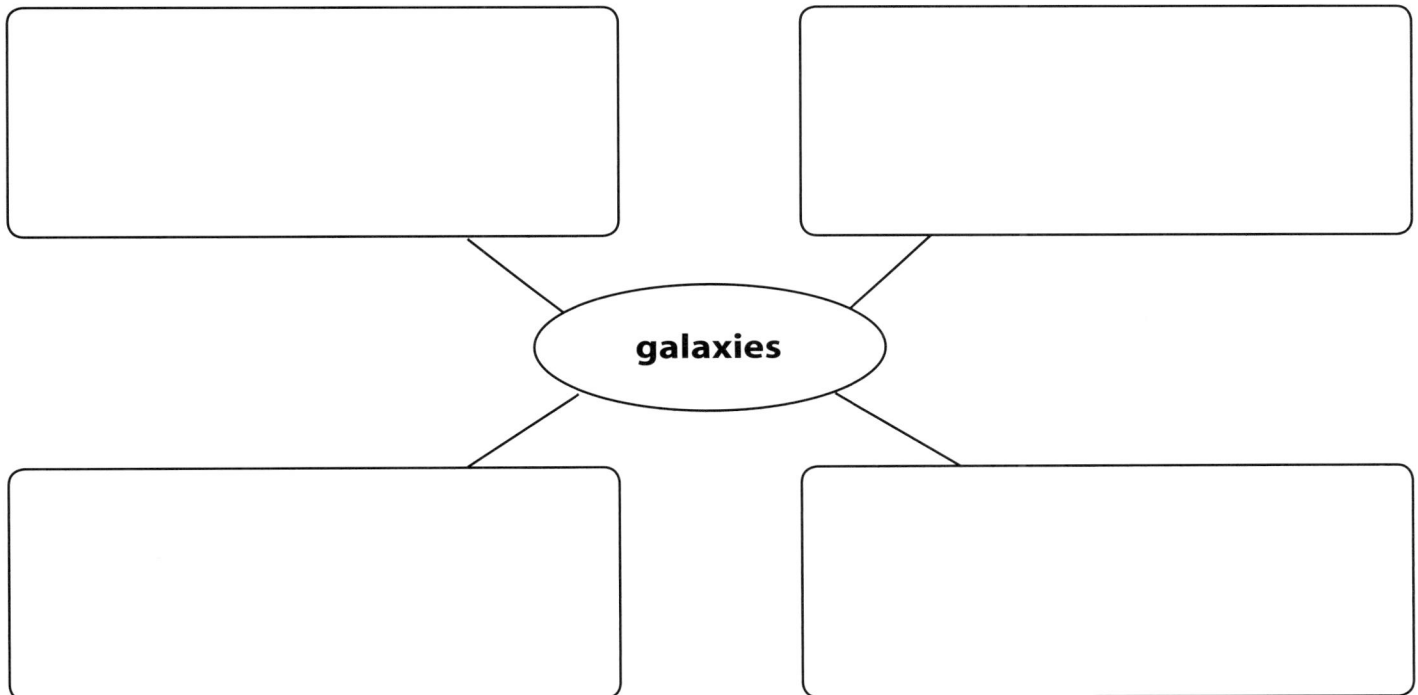

galaxies

STRATEGY FOCUS: ASK QUESTIONS

Write one question you have after reading Chapter 3.

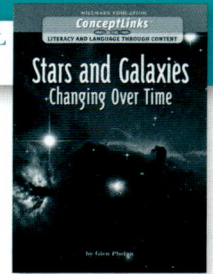

ConceptLinks
Stars and Galaxies
-Changing Over Time

OBJECTIVES

Students will:

LANGUAGE

SPEAK and **LISTEN** to put science terms in order of smallest to largest.

LITERACY

READ and **ANALYZE** nonfiction texts.

INTERPRET diagrams.

ASK questions to deepen understanding.

CONTENT

UNDERSTAND that every star belongs to a galaxy, which is classified by its shape.

UNDERSTAND that our sun is part of the Milky Way Galaxy.

MATERIALS

- **Student Book**, Orange Level
- **Learning Master 15**
- **Audio CD**, Track 17

FOR MORE SUPPORT

PAGE 18 Point out the word **local** in the third paragraph. Explain that **local** refers to a particular place that is part of a whole. The Local Group of galaxies is a particular cluster of galaxies within the universe. Give other examples of the use of **local**, such as local schools, local people, and local customs. Discuss the meanings with students.

Guide Reading

Activity Overview

Orange Level	small group	5-10 minutes

- Use these leveled notes to guide the reading of Chapter 3 in the Orange Level **Student Book**.
- Have the other book-level small groups work on their own Chapter 3 activities. See the detailed grouping plan on page 42.
- When you leave the group, remind students to complete **Learning Master 15** and the Your Turn activities on page 19 of the **Student Book**.
- Return to page 43 of this guide for answers to the Your Turn activities and additional teaching suggestions.

Read Chapter 3: Groups of Stars

Pages 16-17 Read the title and definitions. Have a student read the Key Ideas. Draw students' attention to the three photos. Write the names of the three kinds of galaxies on the board, and ask students to provide details about each kind. Ask: **Where do new stars form in a spiral galaxy? Why?** (in the arms because they contain a lot of gas and dust) Say: **Elliptical galaxies do not have as much gas and dust as spiral galaxies. How old do you think stars are in an elliptical galaxy compared to stars in a spiral galaxy?** (The stars in an elliptical galaxy are older.) **Do you think irregular galaxies have a lot of gas and dust? Why or why not?** (They have a lot of gas and dust because they are made mostly of young stars.)

> **SHARE IDEAS** *Put these terms in order from smallest to largest:* **cluster, star, universe, galaxy**. (star, galaxy, cluster, universe)

Page 18 Read aloud the caption. Discuss the photo with students. Explain that all galaxies are in constant motion. Use a classroom clock or draw a clock on the board, and point out how the hands move clockwise. Help students to see that the center of the galaxy is rotating clockwise, and the arms are trailing behind. Ask: **How long does it take our solar system to revolve one time around the center of the Milky Way Galaxy?** (about 200 million years) Have students recall that a cluster is a group of similar things. Ask: **What is the name of the cluster that the Milky Way Galaxy belongs to?** (the Local Group) Emphasize that our sun is only one star among the billions of stars that make up the Milky Way Galaxy.

Stars and Galaxies: Changing Over Time
Chapter 3: Groups of Stars

ORANGE LEVEL
Student Book,
pages 16-18

USE KEY WORDS

Look at the Key Words on page 23 of your book.
Answer these questions about the Key Words in Chapter 3.

KEY WORDS
galaxies
gravity
universe

1. Groups of stars, gas, and dust are called _____ .

2. _____ holds **galaxies** together.

3. The _____ contains all the **galaxies** in space.

ORGANIZE IDEAS

As you read Chapter 3, complete the chart by writing details about each type of galaxy.

TYPES OF GALAXIES

Spiral Galaxy	Elliptical Galaxy	Irregular Galaxy
1.	1.	1.
2.	2.	2.
3.	3.	3.

STRATEGY FOCUS: ASK QUESTIONS

What is one question you still have about galaxies?

_____

Where could you look to find the answer to your question?

_____

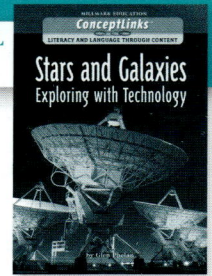

EXPLORE CHAPTER 3 *(cont'd)*

PURPLE LEVEL
pages 16-18

ConceptLinks
Stars and Galaxies
Exploring with Technology

OBJECTIVES

Students will:

LANGUAGE

SPEAK and **LISTEN** to share ideas about exploring the universe.

LITERACY

READ and **ANALYZE** nonfiction texts.

USE visuals to aid comprehension.

ASK questions to deepen understanding.

CONTENT

UNDERSTAND that the big bang theory states that the universe began with an explosion and has been expanding ever since.

MATERIALS

- **Student Book**, Purple Level
- **Learning Master 16**
- **Audio CD**, Track 21

FOR MORE SUPPORT

PAGE 16 Read aloud the first sentence in the first paragraph. Explain that the word **before** refers to **before space telescopes were invented**. Without space telescopes, astronomers could not explore the universe like they are able to now.

Guide Reading

Activity Overview

| Purple Level | small group | 5-10 minutes |

- Use these leveled notes to guide the reading of Chapter 3 in the Purple Level **Student Book**.
- Have the other book-level small groups work on their own Chapter 3 activities. See the detailed grouping plan on page 42.
- When you leave the group, remind students to complete **Learning Master 16** and the Your Turn activities on page 19 of the **Student Book**.
- Return to page 43 of this guide for answers to the Your Turn activities and additional teaching suggestions.

Read Chapter 3: Exploring the Universe

Page 16 Read aloud the title and the caption. Have students describe what they see in the photo. Ask: **How long ago did these galaxies give off the light captured by the Hubble Space Telescope?** (13 billion years ago) **What might be true about some of the galaxies you see here?** (Possible response: They might have more stars or fewer stars now.) Have students discuss the value of space telescopes.

Page 17 Direct students' attention to the three photos. Have students read each label and caption. Discuss the shape of each galaxy. Have students name other things that have spiral, elliptical, and irregular shapes. Ask: **In which type of galaxy is our sun?** (spiral) Tell students that galaxies are in constant motion. Ask: **How can starlight gathered by the Hubble Space Telescope tell scientists about something that happened in the past?** (It takes many light-years for light from distant stars to reach Earth.)

Page 18 Read aloud the Key Idea. Discuss the big bang theory. Explain that all matter in the universe was in one place before the explosion. Point out the dots on the balloon. Have students explain what happens to the dots as the boy blows air into the balloon. (They move away from one another.) Write **universe** on the board. Discuss what makes up the universe. Ask: **How old do scientists believe the universe is?** (14 billion years)

Stars and Galaxies: Exploring with Technology
Chapter 3: Exploring the Universe

PURPLE LEVEL
Student Book,
pages 16-18

USE KEY WORDS

Look at the Key Words on page 23 of your book.
Answer these questions about the Key Words in Chapter 3.

1. The _____ is an idea that explains how the **universe** began.

2. Billions of _____ exist in the **universe**.

KEY WORDS

big bang theory
galaxies
universe

ORGANIZE IDEAS

As you read Chapter 3, write science facts about the big bang theory.

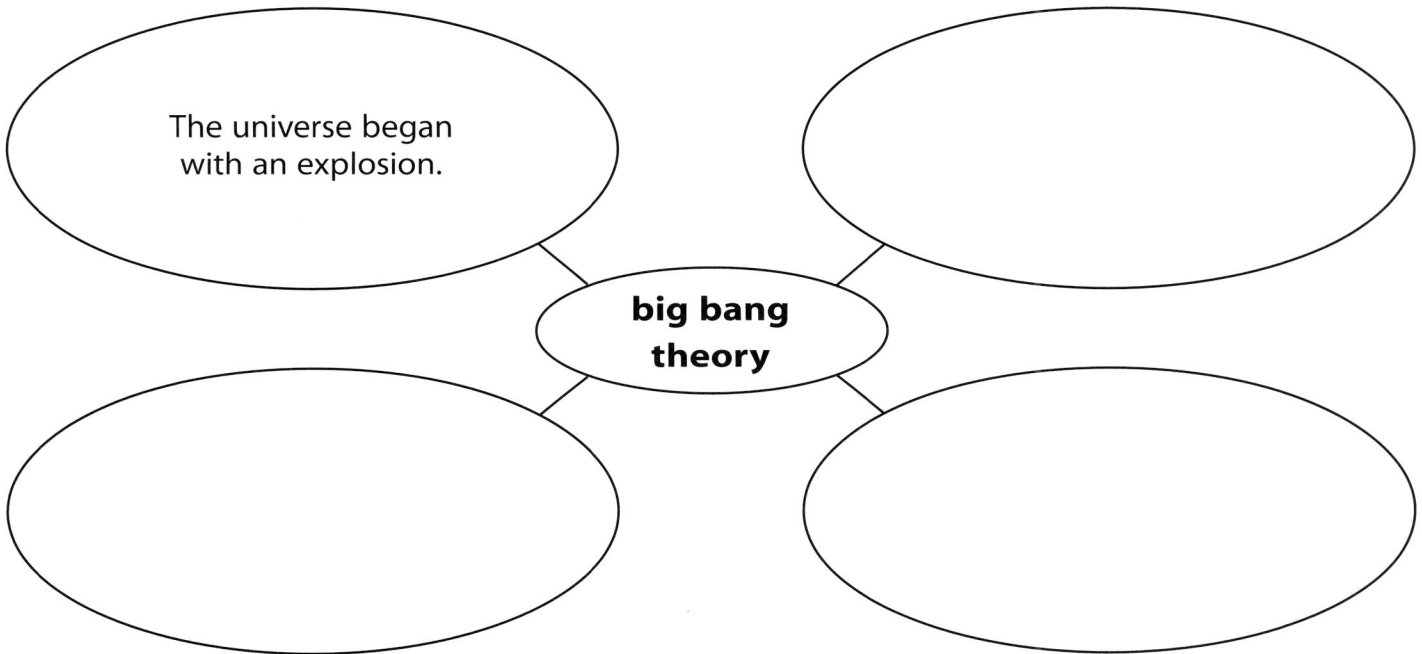

The universe began with an explosion.

big bang theory

STRATEGY FOCUS: ASK QUESTIONS

What is a good source of information to answer any remaining questions you have?
Explain why it would be a good source.

_____

_____

_____

Review Concepts and Vocabulary

15-20 minutes (TG, pp. 52-53)

• Review Concepts

• Review Key Words

Assess and Present

30-60 minutes (TG, pp. 54-61)

• Administer Written Assessments

• Observe Oral Language

• Guide Presentations

OBJECTIVES

Students will:

LANGUAGE

USE academic language to communicate information, ideas, and concepts in science.

LITERACY

MAKE CONNECTIONS from text to world.

INTERPRET charts and diagrams.

CONTENT

DEMONSTRATE knowledge of concepts and vocabulary related to stars and galaxies.

MATERIALS

• **Concept Connector**, Side B

• **Learning Master 17**

• **Image Bank CD-ROM**

• **Resource CD-ROM**

ResourceLinks

See **E-Masters 9**, **9A**, **10**, and **10A** on the **Resource CD-ROM** for printable versions of the **Concept Connector** questions and answers.

Review Concepts and Vocabulary

Activity Overview

Review Concepts	whole group	10 minutes
Review Key Words	whole group	5-10 minutes

Review Concepts

Distribute copies of the Review Guide, **Learning Master 17**. Display the **Concept Connector**, Side B, folding back the Side B flap. You can also use the **Image Bank CD-ROM** to print or project the Side B images. Use the leveled questions on the Side B flap to review concepts related to stars and galaxies. Students can use the Review Guide to label the images and record their responses to the questions.

Answers for Concept Connector, Side B

1. A galaxy is a group of stars, dust, and gases in space.

2. The star in the Big Dipper that is closest to Earth is 65 light-years away.

3. The Milky Way Galaxy is shaped like a big disk or spiral, with a flat area surrounding a ball-shaped area in the middle.

4. Gravity is important for galaxies because it holds the stars, dust, and gases together.

5. Apparent magnitude measures how bright a star looks from Earth, and absolute magnitude measures how bright the star really is.

6. Fusion is important in the life cycle of a star because it supplies the energy in a star. A star goes through different stages in its life cycle as fusion uses up the atoms in the star.

7. If both stars appear to have the same brightness from Earth, the star that is farther away has greater absolute magnitude.

8. The Hubble Space Telescope is in space, which means air, light, and pollution on Earth do not interfere with the light it collects from objects in space.

Review Key Words

For an oral vocabulary review, read aloud each of the Key Words listed on the Side B flap. For each word, have students find an example in the images or make up a sentence that uses the word. Students can record their sentences on the Key Words list, **E-Master 6**.

NAME

Stars and Galaxies

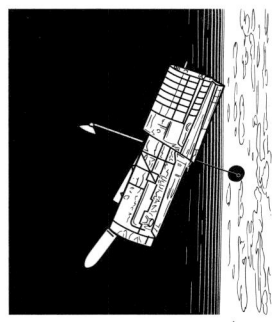

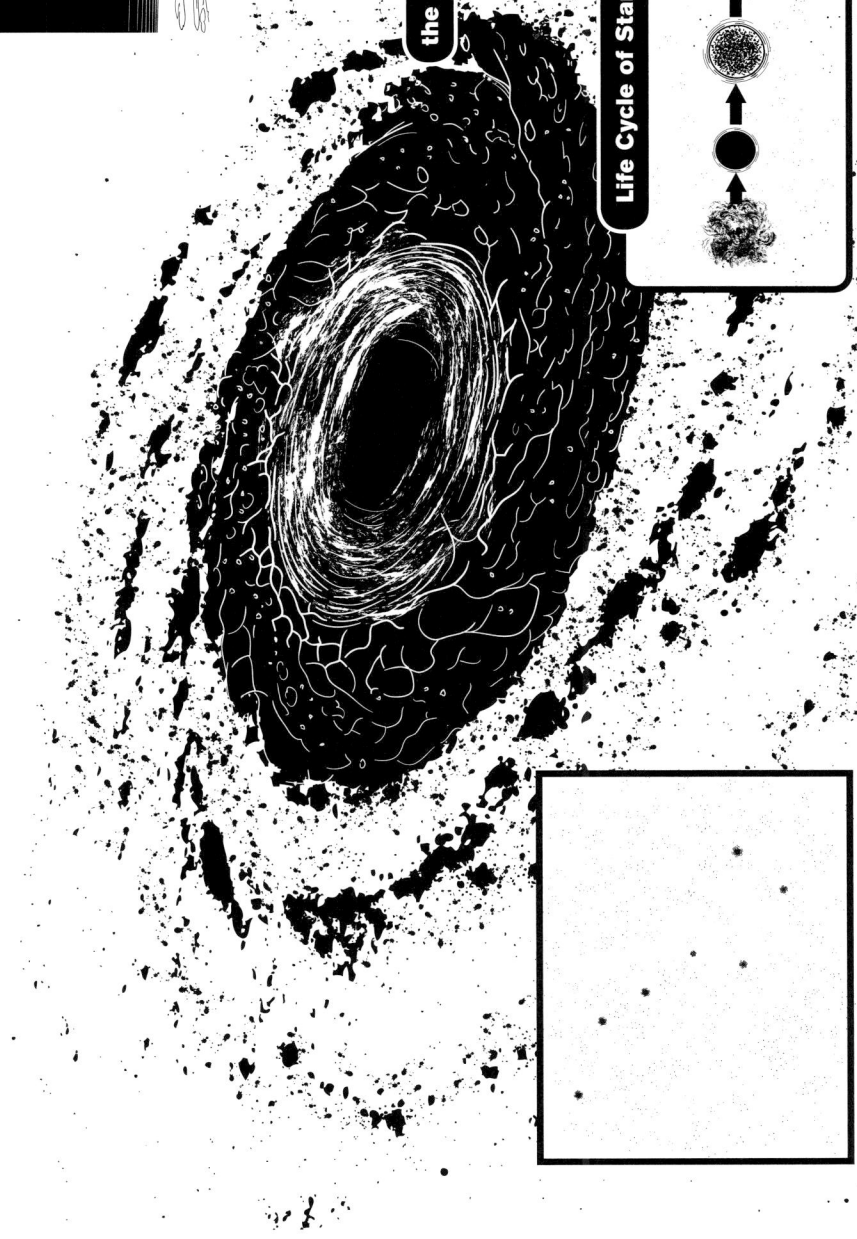

the Milky Way Galaxy

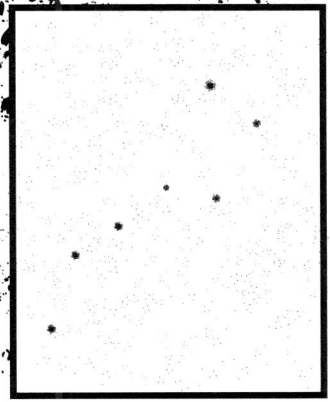

Life Cycle of Stars of High Mass

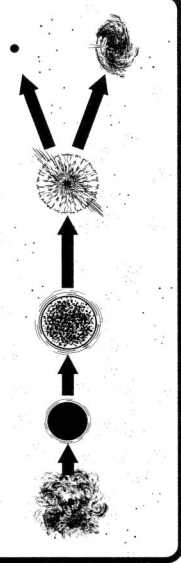

OBJECTIVES

Students will:

LANGUAGE

USE language to ask questions, compare, show sequence, or explain.

EXPLORE academic vocabulary.

LITERACY

APPLY the comprehension strategy of asking questions.

SELF-ASSESS and **REFLECT** on a literacy experience.

CONTENT

DEMONSTRATE knowledge of concepts and vocabulary related to stars and galaxies.

MATERIALS

- **Learning Masters 3, 17-23**
- **Concept Connector**, Side B
- **Student Books**, all levels
- **Resource CD-ROM**
- **Image Bank CD-ROM**

ResourceLinks

Use **E-Masters 1** and **2** on the **Classroom Management and Assessment CD-ROM** to plan and monitor assessment activities.

Assess and Present

Activity Overview

Administer Written Assessments	individuals	20-30 minutes
Observe Oral Language	individuals, partners, small groups	
Guide Presentations	individuals, partners, whole group	10-30 minutes

Administer Written Assessments

Choose from the following written assessment tools to measure your students' progress:

- **Self-Assessment** Distribute copies of the Self-Assessment, **Learning Master 18**. To complete the guide, students should refer to the word lists on inside front covers of the **Student Books** they have read.
- **End-of-Book Tests** Distribute copies of the appropriate End-of-Book Tests, **Learning Masters 19-22**. Go through the directions for each test to make sure students understand the tasks. To provide more support, you can: read the test aloud; have students refer to their Word Maps (**Learning Master 3**) and Review Guides (**Learning Master 17**); let them complete the test with a partner; or make it an open-book test. For answers, see page 87 of this guide.
- **Comprehension Strategy Assessment** Distribute copies of the Strategy Assessment, **Learning Master 23**. Have students read the passage and answer the questions. If students have difficulty reading the passage, read it aloud or allow them to read with a partner. For answers, see page 88 of this guide.

Observe Oral Language

While students are completing the written assessments, you may wish to evaluate their progress with the language focus of asking questions, comparing, showing sequence, or explaining. Work with each book-level group, with partners, or with one student at a time. Have students discuss the light-year chart, diagram of a star's life cycle, and photos of the Milky Way Galaxy and Hubble telescope on the **Concept Connector**, Side B. Use the chart below to assign a specific task, and note whether students incorporate any of the signal words in their conversations. For an Oral Language Rubric, see page 83 of this guide.

	BLUE LEVEL	GREEN LEVEL	ORANGE LEVEL	PURPLE LEVEL
SPEAKING FOCUS	**Ask Questions**	**Compare**	**Show Sequence**	**Explain**
TASK	Have students **ask questions** about the Milky Way Galaxy.	Have students **compare** two stages in the life cycle of a high-mass star.	Have students **show sequence** to explain the stages in the life cycle of a high-mass star.	Have students **explain** why stars that are different distances from Earth have the same apparent magnitude.
SIGNAL WORDS	*how far is, how large is, why does, why is*	*brighter than, hotter than, farther than*	*first, then, next, during, after that, finally*	*as near as, as far as, as bright as, as small as*

Guide Presentations

Students can prepare and orally present the results of the following activities, introduced earlier in the lesson sequence. If feasible, allow students to choose the activity they will prepare and present. To evaluate these activities, refer to the Oral Language Rubric on page 83.

Use the Language of Science (**Student Books**, page 9)

Students can use the academic language that is modeled on page 9 of the **Student Books** as the basis for a longer conversation about the Key Ideas. Students can refer to the Key Ideas maps on the inside back covers of the **Student Books** to compose other questions and answers. Possible formats:

- a mock interview between a journalist and a scientist about the Milky Way Galaxy
- a bilingual presentation of questions and answers about the Key Ideas (For home language resources, see the **Resource CD-ROM**.)
- a conversation between two scientists in the field, observing photos taken by the Hubble Space Telescope of the Crab Nebula

Expand Vocabulary (**Student Books**, page 19)

Students can present their responses to the Expand Vocabulary activities on the Chapter 3 Your Turn pages. Encourage them to support their presentations with visuals, such as images from the **Image Bank CD-ROM** and their own drawings.

Reinforce and Extend: Launch Projects (**Student Books**, pages 20-22)

If students have already carried out the Career Explorations, Use Language/ Writer's Workshop, or Science Around You activities in the **Student Books**, invite them to share their results. See the teaching notes for these activities on pages 62-63 of this guide.

> **Lesson Wrap-up** Encourage students to give each other feedback on their presentations, telling what they noticed, what they learned, and what they liked best.

HOME LANGUAGE SUPPORT

Students may benefit from giving bilingual presentations. For example, one partner can speak in his or her home language, and the other can restate the information in English. Encourage students to consistently use one language at a time.

ResourceLinks

Allow students to use the **Image Bank**, Expand Vocabulary, to find visual support for the activities on page 19 of the **Student Books**.

Book Title _____   **ALL LEVELS**

Look at the inside front cover of your book.

Think about the Science Vocabulary Words. How well do you know these words now?

Write each word in one of the boxes below.

SCIENCE VOCABULARY WORDS		
I already knew these words.	**I learned more about these words.**	**I am still not sure about the meaning of these words.**

Tell something you learned about stars and galaxies.

I learned that _____

_____ .

What was difficult for you to understand?

It was difficult for me to understand _____

_____ .

What activities did you like the best?

I liked _____

_____ .

What questions do you still have about stars and galaxies?

_____

_____

What would you like to read about next?

I would like to read about _____

_____ .

Stars and Galaxies: Light from Far Away
End-of-Book Test

BLUE LEVEL

MULTIPLE CHOICE

Read each item. Select the best answer.

1 POINT EACH

1. A pattern of stars is called a _____.

   A. group   **B.** light-year   **C.** constellation   **D.** galaxy

2. Which is the closest star to Earth?

   A. Barnard's Star   **B.** Sirius   **C.** Proxima Centauri   **D.** the sun

3. Which object reflects light?

   A. the sun   **B.** a star   **C.** a light-year   **D.** the moon

4. Which force causes galaxies to form?

   A. dust   **B.** gravity   **C.** stars   **D.** gases

YES OR NO

Read each item. Write **YES** if it is a star. Write **NO** if it is not a star.

1 POINT EACH

5. the sun _____

6. Earth _____

7. the moon _____

8. Sirius _____

COMPLETE THE SENTENCE

Circle the correct word to complete each sentence.

1 POINT EACH

9. Scientists measure long distances in space in (light-years, constellations).

10. Every star belongs to a (constellation, galaxy).

Stars and Galaxies: Our Universe
End-of-Book Test

MULTIPLE CHOICE

Read each item. Select the best answer.   **1 POINT EACH**

1. Which makes its own light?

   A. a planet   **B.** a star   **C.** Earth   **D.** the moon

2. The hottest stars are _____.

   A. blue   **B.** yellow   **C.** red   **D.** white

3. The shape of the Milky Way Galaxy is _____.

   A. elliptical   **B.** round   **C.** irregular   **D.** spiral

FILL IN THE BLANK

Read the items. Write the missing words in the blanks.   **1 POINT EACH**

4. Scientists use _____ to collect light from stars.

5. _____ is how bright a star really is.

6. The _____ has billions of galaxies.

7. _____ in a star combine and release energy.

8. A small star might have a high _____ because it is close to Earth.

SHORT ANSWER

Read each question. Write an answer in 1-2 complete sentences.   **1 POINT EACH**

9. Why do scientists use light-years to measure distances to most stars?

_____

_____

_____

10. Compared to other stars, how big is the sun?

_____

_____

_____

Stars and Galaxies: Changing Over Time
End-of-Book Test

ORANGE LEVEL

MULTIPLE CHOICE

Read each item. Select the best answer.

1 POINT EACH

1. Which causes a star to give off energy?

   A. a nebula   **B.** gravity   **C.** fusion   **D.** a light-year

2. How long a star lives depends on its _____.

   A. nebula   **B.** brightness   **C.** color   **D.** mass

3. Which is part of the life cycle of a high-mass star?

   A. a white dwarf   **C.** a red giant

   B. a supernova   **D.** a planetary nebula

4. What color star is the sun?

   A. white   **B.** orange   **C.** red   **D.** yellow

5. A galaxy that has the shape of a flattened ball is _____ galaxy.

   A. a cluster   **B.** an irregular   **C.** an elliptical   **D.** a spiral

6. Our solar system includes everything that moves around the _____.

   A. galaxy   **B.** sun   **C.** cluster   **D.** moon

SHORT ANSWER

Read each question. Write each answer in 2-3 sentences.

2 POINTS EACH

7. What is the difference between apparent magnitude and absolute magnitude?

_____

_____

_____

8. How does a star form inside a nebula? Use the words **gravity** and **fusion** in your description.

_____

_____

_____

Stars and Galaxies: Exploring with Technology
End-of-Book Test

MULTIPLE CHOICE

Read each item. Select the best answer.

1. Different kinds of telescopes collect different kinds of _____.

   A. light   **B.** X-rays   **C.** energy   **D.** atoms

2. The real brightness of a star depends on its _____.

   A. temperature and distance from Earth   **C.** size and distance from Earth

   B. size and temperature   **D.** distance from other stars

3. Which describes an invisible area where gravity pulls in all energy and matter?

   A. a mass   **B.** a neutron star   **C.** a black hole   **D.** a supernova

ORDER THE EVENTS

Number the events in the order they happen in the life cycle of a star of high mass. Write **1** for the first event, **2** for the second event, **3** for the third event, and **4** for the last event.

4. _____ Material becomes a black hole.

5. _____ The star expands into a supergiant.

6. _____ Gravity pulls gas and dust in a nebula into a ball.

7. _____ The star explodes in a supernova.

SHORT ANSWER

Read the item. Write an answer in 2-3 complete sentences.

8. Explain the big bang theory.

   _____

   _____

   _____

   _____

NAME _____

Ask Questions

Solar Storms

ALL LEVELS

Storms on Earth can be rainy, snowy, or windy. But what is a storm like on the sun? A **solar storm** is a huge explosion from the outer part of the sun. A large cloud of gas blasts into space. This **solar flare** contains very hot particles.

Some of these particles reach Earth. We can't see the particles, but we can notice their effects. The particles can damage weather and communications satellites. They can even cause power blackouts.

Near the North and South Poles, the gas particles enter Earth's atmosphere. They collide with gases in the air and cause the sky to glow. These **auroras** happen regularly. But during a solar storm, the auroras are much brighter. They may look like white, green, red, and purple ribbons glowing in the sky.

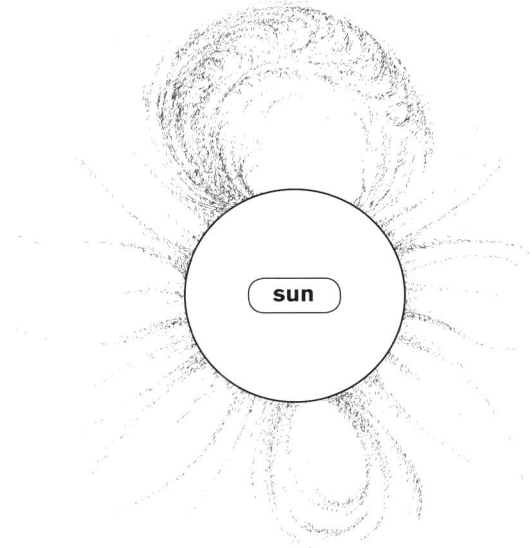

sun

▲ The cloud of particles from a solar storm can grow to be as wide as 48 million kilometers (about 30 million miles) across. This is wider than the sun.

ASSESS THE COMPREHENSION STRATEGY

1. Look at the article title and diagram before you read the article. Write one question you have about solar storms. As you read the article, write another question. Then write one question you still have after reading the article.

BEFORE READING	DURING READING	AFTER READING

2. After reading the article, reread your questions. Were you able to answer any of the questions? Write any answers you found. How can you find answers to the questions you still have?

_____

_____

_____

_____

REINFORCE AND EXTEND

OPTIONAL ACTIVITIES

Launch Projects
5-180 minutes (TG, pp. 62-63)
- Career Explorations
- Use Language/Writer's Workshop
- Science Around You

Assign Further Reading
30-60 minutes (TG, pp. 64-65)
- Select Another Student Book
- Read Another Student Book

Teach Concepts with Visual Mini-lessons
5-15 minutes per mini-lesson (TG, pp. 66-68)
- Plan a Visual Mini-lesson
- Teach the Mini-lesson

OBJECTIVES

Students will:

LANGUAGE
SPEAK and **LISTEN** to share research findings.

LITERACY
USE Internet and print resources to gather information.

CONTENT
EXPLORE careers related to stars and galaxies.

MATERIALS

- **Student Books**, all levels
- **Learning Master 33**
- Print and/or Internet resources on careers
- **Image Bank CD-ROM**

If... you want to extend students' learning with projects related to their reading,

then... use the suggestions below to have students carry out one or more of the Student Book projects.

Launch Projects

Activity Overview

Career Explorations	partners, small groups	10-45 minutes
Use Language/Writer's Workshop	small groups, partners, individuals	20-120 minutes
Science Around You	partners, individuals	5-15 minutes

Career Explorations

The Career Explorations features on page 20 of each leveled **Student Book** present a variety of careers related to stars and galaxies. To research these careers, students can do the following with others reading the same **Student Book** or a different one:

- Read the Career Explorations feature on page 20 of the **Student Book**.
- Use the Internet or print resources to learn more about the featured careers. See www.millmarkeducation.com/conceptlinks/starsandgalaxies for links to career-related websites.
- Allow students to search the **Image Bank** for photos of the featured careers.
- Share the results of their research with a larger group. Possible formats:
  - ▶ an oral report on one or more careers
  - ▶ an interview with one student in the role of a professional in the field they have investigated
  - ▶ a poster with pictures, labels, and captions about one career
  - ▶ a computer-generated presentation on one or more careers

For an Oral Language Rubric, see **Learning Master 33** on page 83 of this guide.

BLUE LEVEL	GREEN LEVEL	ORANGE LEVEL	PURPLE LEVEL
Students learn what software engineers do and tell whether they would like to build software programs.	Students read about a science magazine editor to learn what a career in this field is like and what they might need to prepare for the career.	Students learn what astronauts do and name one exciting thing they would like to do if they were an astronaut.	Students read the schedule of a public relations officer and explain why this work does or does not sound interesting to them.

Use Language/Writer's Workshop

The Use Language activities on page 21 of each **Student Book** help students recognize and use different functions of language when speaking, listening, reading, and writing. Work with each book-level group to introduce and practice the target language function for that level.

SPEAKING FOCUS Read aloud the top part of the page and discuss the examples with students. Then pair students and have them generate other sentences that use language in the same way.

WRITING FOCUS Read aloud the writing prompt and go through the list of Words You Can Use. Give students time to research their topics before writing. Invite them to find and use related images on the **Image Bank CD-ROM**.

For an extended writing process, see the Writer's Workshop and differentiated Writing Guides on pages 76-81 of this guide. Also see the Writing Rubric on page 82.

	BLUE LEVEL	GREEN LEVEL	ORANGE LEVEL	PURPLE LEVEL
SPEAKING FOCUS	Ask Questions	Compare	Show Sequence	Explain
SIGNAL WORDS	how does, how do, how is, how far is, how large is, why does, why do, why is	but, even, much, brighter than, farther than, hotter than	first, then, next, later, before, during, after, begin, become, stop	as near as, as far as, as big as, as small as, as bright as, as dim as
WRITING FOCUS	Write questions and answers about a star.	Write a comparison of two stars in the Milky Way Galaxy.	Use sequence words to tell the life cycle of a star.	Explain the features of a newly-discovered star using examples.

Science Around You

The Science Around You features on page 22 of each **Student Book** help students develop visual literacy and apply academic concepts to real-world contexts. Students can work alone or with partners to do the following:

- Read the Science Around You feature on page 22 of the **Student Books**.
- Carry out the response activity with a partner or in a small group.
- Share their responses with a larger group.

BLUE LEVEL	GREEN LEVEL	ORANGE LEVEL	PURPLE LEVEL
Students read a poster and answer questions about a sky watch.	Students analyze a star map to answer questions about constellations.	Students read a science article and answer questions about the star Fomalhaut and its planets.	Students study a poster to answer questions about the images of a supernova taken by different kinds of telescopes.

OBJECTIVES

Students will:

LANGUAGE — **USE** language to ask questions, compare, show sequence, or explain.

LITERACY — **LOCATE**, **ORGANIZE**, and **INTERPRET** written information.

WRITE, **REVISE**, and **EDIT** texts for specific purposes.

CONTENT — **COMMUNICATE** information about stars and galaxies.

MATERIALS

- **Student Books**, all levels
- **Learning Masters 28-32** (optional)
- **Image Bank CD-ROM**
- Print and/or Internet resources

OBJECTIVES

Students will:

LANGUAGE — **SPEAK** and **LISTEN** to communicate information and ideas.

LITERACY — **CONSTRUCT** meaning from a variety of informational materials.

CONTENT — **APPLY** science concepts to real-world contexts.

MATERIALS

- **Student Books**, all levels

ALL LEVELS

OBJECTIVES

Students will:

LANGUAGE

USE academic vocabulary in new contexts.

LITERACY

APPLY background knowledge from previous reading to a new text.

CONTENT

EXPAND knowledge of concepts related to stars and galaxies.

MATERIALS

- **Student Books**, all levels
- **Learning Masters 1-22**
- **Audio CD**
- **Image Bank CD-ROM**
- **Resource CD-ROM**

ResourceLinks

See the **E-Master** section of the **Resource CD-ROM** for printable versions of the **Concept Connector**, Side A, questions and answers. Also see **E-Masters 6**, **7**, and **11-18** for printable versions of the Key Words and Key Ideas. Have students use these **E-Masters** for notes or home language support.

If... you want students to build on their learning by reading more about stars and galaxies,

then... use the suggestions below to have students read another *ConceptLinks™* Stars and Galaxies book.

Assign Further Reading

Activity Overview

Select Another Student Book	teacher planning time	
Read Another Student Book	individuals, partners, small groups	30-60 minutes

Select Another Student Book

Students can deepen their knowledge of concepts and vocabulary by reading another *ConceptLinks™* Stars and Galaxies book. Consider these options for placing students in another book:

- Students who have read the Blue, Green, or Orange Level books can read a higher-level book with a more proficient partner, with the **Audio CD**, or in a teacher-guided small group.
- Students who have read the Orange or Purple Level books can read a lower-level book alone for fluency practice, or with a less proficient partner for a cooperative learning experience.

Read Another Student Book

To guide the reading, use the lessons on pages 14-55 of this guide with some modifications. See the suggestions below for adapting each lesson and the Partner Reading Strategies on page 65.

Preview

- **Introduce Stars and Galaxies (pages 14-15)** Revisit the **Concept Connector**, Side A, and have students share new understandings of what they see in the photos. Students at the earliest levels of language proficiency can add more labels and captions to their copy of **Learning Master 1**. Other students can do a three-minute quickwrite by responding to this question: **What is the process in which stars make their own light?**
- **Prepare for Reading (pages 16-17)** As students preview the new book, they can use a chart such as the one shown here to write questions and then record the answers as they find them.

My Questions About Stars and Galaxies	My Answers About Stars and Galaxies
What is a supernova?	the bright explosion of a dying supergiant star

- **Build Academic Vocabulary (pages 18-19)** Have students discuss additional information that they can add to their Word Maps (**Learning Master 3**).

Explore Chapter 1

- **Develop Strategic Learners (pages 20-21)** First, briefly recap the comprehension skill of Asking Questions. Then have students read the Chapter 1 title and note any questions they have about the title. Ask what questions they have about the image on the page.
- **Read Chapter 1 (pages 22-31)** To provide scaffolding for students who are reading at a more challenging level, see For More Support in the Guide Reading lessons.

Explore Chapter 2

- **Read Chapter 2 (pages 32-41)** To scaffold the content, do a detailed preview of the chapter, discussing the headings, pictures, captions, charts, and diagrams on each page. Also see For More Support in the Guide Reading lessons.

Explore Chapter 3

- **Read Chapter 3 (pages 42-51)** To scaffold the content, review the Key Ideas from previous chapters. Also see For More Support in the Guide Reading lessons.

Wrap Up

- **Review Concepts and Vocabulary (pages 52-53)** Have students make up their own questions and answers about the Milky Way Galaxy shown on the **Concept Connector**, Side B. Invite students to identify some Key Words they now feel they know very well and some Key Words they still do not know very well.
- **Assess and Present (pages 54-55)** If the End-of-Book Test at the new level is very difficult for students, read the test items aloud. Alternately, make the test an open-book test, and observe how well students can locate the information. To assess the comprehension strategy again, choose a nonfiction passage from other classroom materials and adapt the questions on **Learning Master 23** for the new passage.

PARTNER READING STRATEGIES

- As students work together, have them take turns reading aloud, each reading one page at a time.
- Alternately, a more proficient reader can read aloud each page first, and the less proficient reader can do an echo reading.
- Have students work together to complete the Study Guide **Learning Masters**.
- When students have completed the reading, invite them to retell the chapter to each other or another set of partners.

ResourceLinks

See the **E-Master** section of the **Resource CD-ROM** for printable versions of the **Concept Connector**, Side B, questions and answers. Have students use these **E-Masters** for notes or home language support.

ALL LEVELS

OBJECTIVES

Students will:

LANGUAGE

SPEAK and **LISTEN** to share ideas about stars and galaxies.

LITERACY

USE visuals to aid comprehension.

CONTENT

INTERPRET charts and diagrams to show understanding of science concepts.

MATERIALS

- **Student Books**, all levels
- **Image Bank CD-ROM**
- **Resource CD-ROM**

ResourceLinks

See the Mini-lesson Planning section of the **Classroom Management and Assessment CD-ROM** for the Mini-lesson planner, **E-Master 19**, and the maps, charts, and diagrams.

If... you want to use visual aids to introduce or review concepts related to stars and galaxies,

then... use the suggestions below to present visual mini-lessons.

Teach Concepts with Visual Mini-lessons

Activity Overview

Plan a Visual Mini-lesson	teacher planning time	
Teach the Mini-lesson	whole group, small groups	5-15 minutes per mini-lesson

Plan a Visual Mini-lesson

Many of the charts, diagrams, and other visual aids found in the **Student Books** are provided as downloadable files on the **Image Bank CD-ROM** and the **Resource CD-ROM**. These images offer strong visual support for important concepts related to stars and galaxies. You can use these visual aids to:

- introduce new concepts to students;
- review and reteach concepts;
- reinforce and extend academic vocabulary;
- prepare students for high-stakes testing by having them interpret graphic information.

Use the following process to plan visual mini-lessons that reinforce your instructional goals:

1 Identify your language, literacy, and/or content objectives.

2 Choose the Stars and Galaxies visual aids that will support your objectives. Refer to the Visual Aids chart on page 67.

3 Identify the background information students will need. Decide what concepts students should review in order to understand the new concepts.

4 Plan the mini-lesson. Refer to the Mini-lesson Planner, **E-Master 19**, on the **Resource CD-ROM**.

Teach the Mini-lesson

Use the Mini-lesson Planner you have prepared to do the following:

- **Preview** Tell students what they will be learning about.
- **Review** Go over the concepts that will help students understand the new concept.
- **Present & Discuss** Display the visual aid. Guide students through the image, making connections with what they already know and presenting new information. Ask questions to engage their thinking about the concepts and vocabulary.
- **Wrap Up** Conclude the mini-lesson by having students demonstrate their understanding of the concept, either by reflecting on this visual aid or applying their knowledge to another example.

DISPLAY TIPS

To display images for your mini-lesson, use the **Image Bank** to:

- print out copies of the images;
- make transparencies;
- project the images using available technology.

You may also have students refer to the images in the **Student Books**, sharing copies if necessary.

Stars and Galaxies Visual Aids			
Concept/Visual Aids	Image Bank	Location of Image in Student Book	Teaching Support in this Guide
star properties			
• star size diagram	✓	GREEN, page 14	page 36
• stars' surface temperatures chart	✓	GREEN, page 11	page 36
	✓	ORANGE, page 8	page 28
	✓	PURPLE, page 11	page 40
• life cycles of stars diagram	✓	ORANGE, pages 12-13	page 38
	✓	PURPLE, page 13	page 40
distances from Earth			
• stars' distances from Earth chart	✓	BLUE, page 13	page 34
• stars' distances from Earth diagram	✓	GREEN, pages 6-7	page 26
• distances to stars diagram	✓	ORANGE, page 6	page 28
• the Big Dipper diagram	✓	ORANGE, page 7	page 28
exploring space			
• telescopes in space chart	✓	PURPLE, page 8	page 30
• types of galaxies photos	✓	GREEN, page 16	page 46
	✓	ORANGE, pages 16-17	page 48
	✓	PURPLE, page 17	page 50

Instructions: Copy this page, and plan your own mini-lesson on the copy, using the Sample Mini-lesson Planner as a guide. After writing the mini-lesson, you can cut it out for convenient use while presenting the images and teaching the mini-lesson. For an electronic version, see **E-Master 19** on the **Resource CD-ROM**.

SAMPLE MINI-LESSON PLANNER	MINI-LESSON PLANNER
Science Concept: **Life Cycles of Stars**	Science Concept: _____
Objectives	**Objectives**
Students will:	Students will:
• speak and listen to share ideas about the life cycle of stars. (Language)	
• use visuals to aid comprehension. (Literacy)	
• interpret diagrams of star life cycles and temperatures to show understanding. (Content)	
Visual(s)	**Visual(s)**
• life cycles of stars diagram (ORANGE, pages 12-13)	
• stars' surface temperatures chart (ORANGE, page 8)	
Preview	**Preview**
Tell students they will learn how the properties of a star change as it goes through its life cycle.	
Review	**Review**
Terms: nebula, supernova, supergiant, red giant, white dwarf, neutron star, black hole	
Stars vary in color, brightness, and size. The properties of stars are related. For example, the color of a star depends on its surface temperature.	
Present & Discuss	**Present & Discuss**
• Display the life cycle diagram. Explain that the properties of a star change as the star goes through its life cycle. Discuss each stage in the life cycle of a high-mass star and an average-mass star.	
• Show students the stars' surface temperature chart. Have them identify the temperatures of the red giant and the white dwarf in the life cycles diagram.	
• Display objects of different sizes, such as a basketball, softball, golf ball, and gumball. Have students decide which stage in the life cycle of a star each object represents and place them in order. Then ask them to describe the properties of the star in each stage.	
Wrap Up	**Wrap Up**
Have students use color markers to make their own drawings of the life cycle of a high-mass or an average-mass star. For each stage, they should include the name of the the stage and a caption that briefly describes the stage.	

ConceptLinks™

LITERACY AND LANGUAGE THROUGH CONTENT

Additional Resources

Newcomer Lesson 70

Build Concepts and Vocabulary (Learning Master 24) 72

Write About Stars and Galaxies (Learning Master 25) 73

Explore the Student Book (Learning Master 26) 74

Build a Story (Learning Master 27) 75

Writer's Workshop 76

Writing Guide BLUE LEVEL 78

Writing Guide GREEN LEVEL 79

Writing Guide ORANGE LEVEL 80

Writing Guide PURPLE LEVEL 81

Writing Rubric 82

Oral Language Rubric 83

Answers to the Learning Masters 84

Acknowledgments 88

LESSON ACTIVITIES

Starting Up with Newcomers
30-90 minutes (TG, pp. 70-75)
- Introduce Vocabulary
- Build Language and Concepts
- Reinforce and Extend

OBJECTIVES

Students will:

LANGUAGE

USE appropriate discourse patterns to discuss stars and galaxies.

LITERACY

ASSESS prior knowledge of academic vocabulary.

PREVIEW text to explore vocabulary and prepare for reading.

CONTENT

UNDERSTAND that a star is a huge ball of gases.

UNDERSTAND that groups of stars form patterns in the sky.

MATERIALS

- **Student Book**, Blue Level
- **Learning Masters 24-27**
- **Audio CD**, Tracks 1-5, 6-9
- **Image Bank CD-ROM**
- **Resource CD-ROM**

HOME LANGUAGE SUPPORT

For home language support of the Basic Vocabulary, see the **Resource CD-ROM**.

If... you have students with very limited English proficiency,

then... use these teaching notes to introduce basic vocabulary, concepts, and language patterns related to stars and galaxies.

Starting Up with Newcomers

Activity Overview

Introduce Vocabulary	individuals, small group	10-20 minutes
Build Language and Concepts	individuals, partners	20-40 minutes
Reinforce and Extend	individuals, partners	15-30 minutes (optional)

Introduce Vocabulary

Teach basic vocabulary about stars and galaxies to students with very limited English language skills. Use the Word-Learning Strategies below to introduce words from the Blue Level **Student Book**.

Word-Learning Strategies

- Use the Starting Up section of the **Image Bank CD-ROM** to download and print the Stars and Galaxies Basic Vocabulary list and a set of picture cards for each student.
- Say each word. Have students clap out the number of syllables as they repeat each word several times. Have students hold up a picture card that illustrates the word.
- Say a sentence using the word and have students repeat the sentence.
- Have students find a photo or diagram in the Blue Level **Student Book** that shows each word.
- For pronunciation practice, play the **Audio CD**, Track 1, and have students repeat the words while referring to their copies of the Stars and Galaxies Basic Vocabulary list.

STARS AND GALAXIES BASIC VOCABULARY

NOUNS		VERBS		ADJECTIVES	
distance	scientist	belong	shine	bright	far
Earth	sky	explain	travel	closer	huge
heat	space	reflect	use	closest	large
light	star			distant	small
pattern	sun				

Build Language and Concepts

Use the activities on the Starting Up **Learning Masters** to develop language and concepts related to stars and galaxies. Begin each activity by reading the instructions aloud, gesturing for each step. Students can also preview each **Learning Master** with the **Audio CD**. Have them work with partners or more proficient speakers to complete the activities.

Build Concepts and Vocabulary (**Learning Master 24**; **Audio CD**, Track 2)

- Before students begin **Learning Master 24,** use the photos on pages 2-3 of the **Student Book** to talk about stars, using different adjectives including **bright**, **distant**, **huge**, and **hot**.
- Have students look at the adjectives on the **Learning Master** chart. Ask them to name other objects that could be described with these words.

Write About Stars and Galaxies (**Learning Master 25**; **Audio CD**, Track 3)

- Point to the list of nouns. Tell students that a noun names a person, place, or thing. Help students identify each object as you point to the picture.
- Help students make personal connections by asking them to name other things that are bright, dark, distant, large, round, or small.
- Have students share their sentences with a partner.

Explore the Student Book (**Learning Master 26**; **Audio CD**, Track 4)

- Give each student a copy of the Blue Level **Student Book**.
- Show students how to preview the **Student Book** by completing the first item on the **Learning Master** together.

Build a Story (**Learning Master 27**; **Audio CD**, Track 5)

- Read aloud each question and the answer choices. Explain that the pictures support one possible story, but students can choose any details they like for their own stories.
- Have students dictate or write a story based on their answers to the questions. Encourage students to add more details as they build their stories. Have them illustrate their stories.
- Have students retell their stories to a partner or small group.
- Let students know their stories can be humorous, unlikely, or even impossible. Save their stories in a folder for continued sharing.

Reinforce and Extend: Read a Student Book (optional)

When students have completed the **Learning Masters**, you may have them read the Blue Level **Student Book** with you, with a more proficient partner, or with the **Audio CD**, Tracks 6-9. See the Blue Level Guide Reading lessons on pages 24, 34, and 44 of this guide.

> **Lesson Wrap-up** Invite students to read aloud the sentences they wrote in the **Learning Masters** and to share details from their stories.

ResourceLinks

Have students use the Search section of the **Image Bank** to look for visual support for new or difficult words. Also see the **E-Masters** on the **Resource CD-ROM** for support of the Key Vocabulary Words, Key Ideas, and **Concept Connector** questions and answers.

Stars and Galaxies
Build Concepts and Vocabulary

Cut out the pictures. Match each picture to a word.
Glue the pictures to the chart.
Write and say sentences about the pictures.

Words That Describe Stars	
hot	huge
bright	distant

1. The size of a star is _____ .

2. Stars look small in the sky because they are _____ .

3. Stars look _____ because they are _____ .

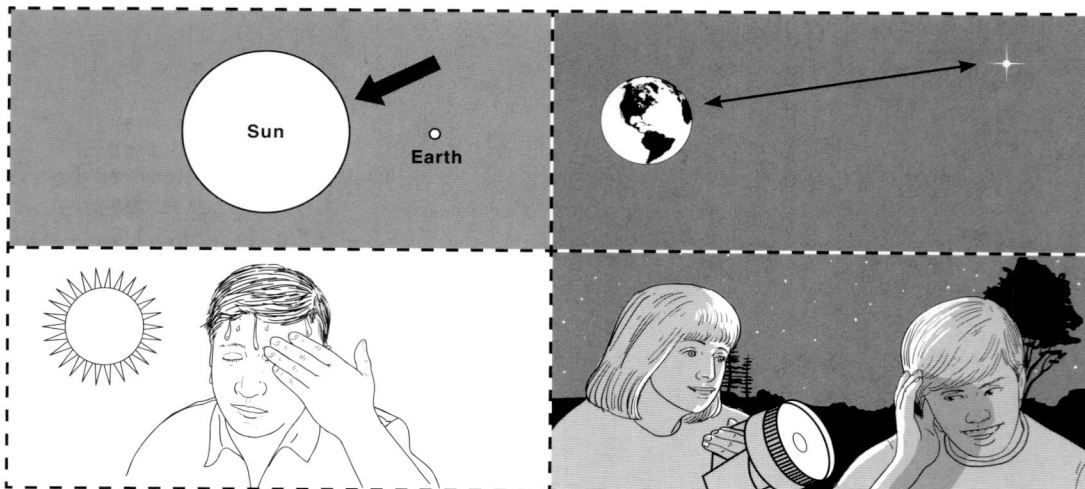

Stars and Galaxies
Write About Stars and Galaxies

Look at the picture. Write sentences about the picture.

Words You Can Use			
Nouns		**Adjectives**	
moon	star	bright	large
scientist	telescope	dark	round
sky		distant	small

1. The scientist looks into a _____.

2. The stars look _____ and _____.

3. The moon looks _____ and _____.

4. The _____ is _____.

5. _____

Stars and Galaxies
Explore the Student Book

Look at *Stars and Galaxies: Light from Far Away.*
Tell what you see on the pages.

1. Look at page 5.

I see the shape of a _____ .

2. Look at page 6.

I see a boy using a _____ .

3. Look at page 10.

I see the _____ . It looks _____ .

4. Look at page 11.

The _____ is much bigger than _____ .

5. Look at page 18.

The light in the sky comes from _____ .

6. Look at another page. Tell what you see.

_____

_____

7. Draw a picture of something interesting from the book.

NAME _____

Stars and Galaxies
Build a Story

Ask each question. Choose an answer to build a story. Tell your story to a partner.

1. You look at stars. What do you use to see them?

a small telescope   a large telescope
just my eyes   binoculars

2. What do you see?

two stars   a small group of stars
many stars   different colored stars

3. What shape do the stars make?
a bear   a lion
a person   a fish

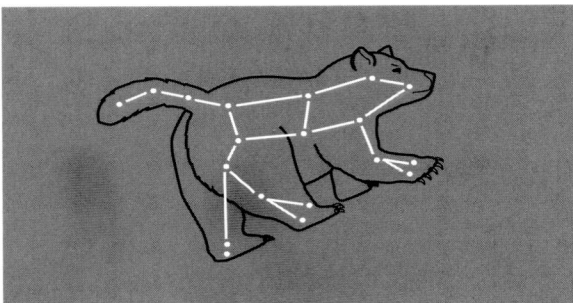

4. What is special about one of the stars?
It is the brightest.   It is red.
It is the largest.   It explodes.

5. What do you do?
I keep looking at the sky.
I read a book about stars.
I draw what I see.
I call a friend.

6. What happens next?
I tell my family to look at the sky.
I write a report about stars.
I draw a map of the night sky.
I call a scientist.

LESSON ACTIVITIES

Use the Writing Process
60-120 minutes
(TG, pp. 76-83)

❶ Prepare for Writing
❷ Draft
❸ Revise
❹ Edit
❺ Publish and Present

OBJECTIVES

Students will:

LANGUAGE

WRITE, **REVISE**, and **EDIT** writing to communicate in science.

LITERACY

USE reference materials.

ORGANIZE information before writing.

CONTENT

APPLY knowledge of stars and galaxies to ask questions, compare, show sequence, or explain.

MATERIALS

- **Student Books**, all levels
- **Learning Masters 3, 28-33**
- **Image Bank CD-ROM**
- **Resource CD-ROM**

HOME LANGUAGE SUPPORT

Students may write more extensively in English if they first use their home language to generate writing ideas.

If... you want your students to follow the writing process, **then...** use these teaching notes to guide the writing process for the prompt on page 21 of each Student Book.

Use the Writing Process

Activity Overview

1. Prepare for Writing	small groups, partners	20-30 minutes
2. Draft	individuals	10-30 minutes
3. Revise	individuals, partners	10-15 minutes
4. Edit	individuals, partners	10-15 minutes
5. Publish and Present	small groups, whole group	10-30 minutes

❶ Prepare for Writing

Have students gather in their book-level small groups. Distribute the appropriate Writing Guide, **Learning Masters 28-31**. Meet with each book-level small group to set up the writing assignment.

- Begin by having students review the writing prompt, which appears on page 21 of the **Student Book** and on the leveled Writing Guide. Then read aloud and discuss the Writing Model.
- To generate ideas, students can refer to their Word Maps (**Learning Master 3**) and to the Writing Ideas on the **Image Bank CD-ROM**. To help students organize their ideas, you can provide a customized graphic organizer from the **Resource CD-ROM**. See details on the chart below.

	BLUE LEVEL	**GREEN LEVEL**	**ORANGE LEVEL**	**PURPLE LEVEL**
WRITING IDEAS Image Bank CD-ROM	Images of various stars	Images of the stars Spica, Adhara, Antares, and Sirius	Images of stars in various phases of their life cycles	Images of various stars with different properties
GRAPHIC ORGANIZERS Resource CD-ROM	Two-column chart for listing questions and answers about a star	Three-column table to compare the features of two stars	Sequence chart showing the life cycle of a star	Idea web to explain details about an imaginary star

If students need to do further research, provide access to reference materials or Internet resources.

❷ Draft

Allow time for students to write a first draft of their assignments. Scaffold the writing according to their proficiency levels:

- Students at lower levels can use the Writing Model as a framework, copying the model and replacing all the underlined words with their own ideas.
- Students at higher proficiency levels can refer to the Writing Model as a guide for their own original writing.

❸ Revise

Have students revise their writing alone or with a Peer Review partner. See the differentiated suggestions for revising on the Writing Guides.

Revising Focus: Synthesis

The following are incremental steps toward producing a text that synthesizes information. Have students incorporate more of the following in their writing as they gain writing proficiency:

- Provide an accurate response to the writing prompt.
- Include information from the research.
- Relate ideas to personal knowledge and experience.
- Present ideas that build upon the science concepts.
- Develop a line of thought that shows reflection and analysis.

❹ Edit

Have students edit their writing alone or with a Peer Review partner. See the differentiated suggestions for editing on the Writing Guides.

Editing Focus: Sentences

Depending on students' levels of writing proficiency, look for these signs of progress in editing sentences:

- Sentences begin with a capital letter and end with a period, question mark, or exclamation point.
- Students understand how to break the flow of text into meaningful sentences.
- Students identify and correct sentence fragments and run-on sentences.

❺ Publish and Present

Allow time for students to prepare a final, clean draft of their work, either by hand or on a computer. Then give students various options for presenting their work:

- as a read-aloud;
- in a display, with accompanying photos or drawings;
- in book form, with a cover;
- as part of a class book on stars and galaxies.

Assess To assess students' writing and oral presentations, see the Writing Rubric and Oral Language Rubric on pages 82-83 of this guide. Students can use the same rubrics to do self-assessments or peer assessments.

PEER REVIEW

Invite partners to read each other's papers aloud and offer a response. Partners should:

- state what they think the paper is about;
- tell what they like about the writing;
- tell something that could be more clear or more detailed;
- point out any words that should be capitalized and any missing end punctuation.

Stars and Galaxies: Light from Far Away
Write Questions and Answers

BLUE LEVEL
Student Book,
page 21

You learned that stars are very large. You learned that stars are very far away. Find out about one star. Write questions and answers about the star.

- Use **how** and **why**.

- Use words from the questions in your answers.

Words You Can Use	
How does...	Why does...
How do...	Why do...
How is...	Why is...
How far is...	
How large is...	

WRITING MODEL

Study the Writing Model. Think about your writing ideas.
Use the Writing Model as an example for your writing.

What is Proxima Centauri?

What is Proxima Centauri?
Proxima Centauri is a bright star in the night sky.
Why is Proxima Centauri so bright?
It is so bright because it is the closest star to Earth after the sun.
How far away is Proxima Centauri from Earth?
Proxima Centauri is about 4 light-years away from Earth.

Revising Focus: Synthesis

- Did you answer the question in the title?
- Did you use words from the questions in your answers?

Editing Focus: Sentences

- Does each sentence begin with a capital letter?
- Does each question end with a question mark?

NAME ______________________________

Stars and Galaxies: Our Universe

Write a Comparison

GREEN LEVEL
Student Book, page 21

Spica, Adhara, Antares, and Sirius are all stars in the Milky Way Galaxy. Choose two of these stars and find out more about them. Write a comparison of the two stars.

- Compare the sizes, colors, and temperatures of the two stars.
- Tell which star is closer to Earth.
- Include other interesting information about the stars.

Words You Can Use

Comparison Words	
but	brighter than
even	farther than
much	hotter than

WRITING MODEL

Study the Writing Model. Think about your writing ideas. Use the Writing Model as an example for your writing.

Spica and Sirius

Spica is the brightest star in the constellation Virgo. Sirius is the brightest star in the constellation Canis Major. Sirius is also the brightest star in the sky. It is 26 times brighter than the sun. But Spica is 2,100 times brighter than the sun!

Why does Sirius look brighter than Spica? Spica is 260 light-years away, and Sirius is only 8.6 light-years away. So Sirius is much closer to Earth than Spica. But both stars are too far away for us to visit them!

Revising Focus: Synthesis

- Did you include information from your research?
- Does your writing use words that compare?

Editing Focus: Sentences

- Does each sentence end with a period, question mark, or exclamation point?

Stars and Galaxies: Changing Over Time
Write to Show Sequence

Look at the diagram on pages 12 and 13. Beginning with the nebula, write a description of the life cycle of a star, from birth to death.

- Decide if you want to describe the life cycle of a star with average mass or high mass.
- Tell about each phase in sequence.
- Use sequence words.

Words You Can Use	
Sequence Words	
first	during
then	after
next	begin
later	become
before	stop

WRITING MODEL

Study the Writing Model. Think about your writing ideas.
Use the Writing Model as an example for your writing.

The Life Cycle of a High-Mass Star

All stars begin life in a nebula. In some places in the nebula, gases and dust are pulled together. They form a huge, hot ball. After the ball gets hot enough, fusion begins. A star is born!

A star with high mass will shine for a few million years. When the star has used up its lightweight atoms in fusion, the star collapses. After it collapses, fusion begins again. The star expands again. Now it becomes a supergiant.

Later, the supergiant collapses. Then it explodes. Now it is called a supernova. After the supernova, a small neutron star or a black hole can form.

Revising Focus: Synthesis

- Does your writing show an understanding of the life cycle of a star?
- Does your writing use sequence words, such as *first*, *then*, and *next*?

Editing Focus: Sentences

- Is it clear where sentences begin and end?
- Is each sentence a complete thought?

Stars and Galaxies: Exploring with Technology
Write to Explain

Imagine you have just discovered a new star.
Write a description of the star.

- Explain what the star looks like and how far away it is.
- Use words that help the reader imagine your new star.
- Compare the star to things most people would be familiar with.

Words You Can Use	
as near as	as small as
as far as	as bright as
as big as	as dim as

WRITING MODEL

Study the Writing Model. Think about your writing ideas.
Use the Writing Model as an example for your writing.

Alpha Mineous

I discovered a new star that I named <u>Alpha Mineous</u>. <u>Alpha Mineous</u> is a very <u>large</u> star. If <u>Earth</u> were <u>the size of a penny, Alpha Mineous</u> would be <u>the size of a hot air balloon.</u>

<u>Alpha Mineous</u> is <u>215</u> light-years away. That means it is <u>50 times as far as Proxima Centauri</u>. When I observe <u>Alpha Mineous</u>, I know the light I see is <u>215</u> years old. When that light left <u>Alpha Mineous, the United States was a new nation. Cars and trains and airplanes had not yet been invented.</u>

<u>Alpha Mineous</u> is <u>a giant blue star</u>, and its color is like <u>the sky on a sunny day</u>. It is <u>much brighter</u> than our sun. The temperature of <u>Alpha Mineous</u> is <u>very hot</u>. <u>It is twice as hot as our sun. Some day, millions of years from now, Alpha Mineous could become a supernova.</u>

Revising Focus: Synthesis

- Does your explanation use information you have learned about stars?
- Does your writing compare the star to things that people are familiar with?

Editing Focus: Sentences

- Have you checked your writing for sentence fragments or run-on sentences?

Assess Writing

You can use this list to think about your own writing or someone else's writing. Look closely at the writing. Decide how well the writing reflects each item on the list. For each item, give a score from 1 to 5.

ALL LEVELS

Characteristics of Good Writing	1 The writing does not do this.	2 The writing does this a little.	3 The writing does this well.	4 The writing does this very well.	5 The writing does this extremely well.
Synthesis*					
• The writing is an accurate response to the writing prompt.	1	2	3	4	5
• The writing includes information from the research.	1	2	3	4	5
Ideas					
• The ideas are clear and focused.	1	2	3	4	5
• The writing shows an understanding of the science concepts.	1	2	3	4	5
Organization					
• The title helps show what the writing is about.	1	2	3	4	5
• The writing uses words that show how different ideas are connected.	1	2	3	4	5
Voice					
• The purpose of the writing is clear.	1	2	3	4	5
• The writing uses words that help: ☐ ask questions. ☐ compare. ☐ show sequence. ☐ explain.	1	2	3	4	5
Word Choice					
• The writing uses some of the Science Vocabulary Words.	1	2	3	4	5
• The writing uses accurate and appropriate words.	1	2	3	4	5
Sentence Fluency					
• The writing has a mix of short sentences and long sentences.	1	2	3	4	5
• The sentences fit together smoothly.	1	2	3	4	5
Conventions					
• Sentences end with a period, a question mark, or an exclamation point.**	1	2	3	4	5
• Words are spelled correctly.	1	2	3	4	5

* Revising Focus **** Editing Focus**

Assess an Oral Presentation

When you give an oral presentation, there are important skills that will help you communicate more effectively.

ALL LEVELS

You can use this list to assess an oral presentation. It can be your presentation or someone else's presentation. Decide how well the presentation reflects each item on the list. For each item, give a score from 1 to 5.

Characteristics of a Good Oral Presentation	1 The speaker did not do this.	2 The speaker did this a little.	3 The speaker did this well.	4 The speaker did this very well.	5 The speaker did this extremely well.
Purpose					
• The purpose of the presentation was clear.	1	2	3	4	5
• The speaker stayed focused on the main points of the presentation.	1	2	3	4	5
Audience					
• The speaker had good eye contact with the audience.	1	2	3	4	5
• The presentation was appropriate for the audience.	1	2	3	4	5
Vocabulary					
• The speaker used a variety of interesting words.	1	2	3	4	5
• The speaker used some of the Science Vocabulary Words.	1	2	3	4	5
Fluency and Pronunciation					
• The speaker spoke smoothly, without long pauses.	1	2	3	4	5
• The speaker pronounced the words clearly enough to be understood.	1	2	3	4	5
Volume and Pacing					
• The speaker spoke loudly enough to be heard by the audience.	1	2	3	4	5
• The pace of the speaker's words was good, not too fast or too slow.	1	2	3	4	5
Visual Support					
• The speaker used gestures and body language to communicate.	1	2	3	4	5
• If visual aids were used, they helped the audience understand the message.	1	2	3	4	5

Stars and Galaxies: Light from Far Away
Chapter 1: Looking at Stars

Study Guide
Learning Master 5

BLUE LEVEL
Student Book,
pages 4–8

USE KEY WORDS

Look at the Key Words on page 23 of your book.
Answer these questions about the Key Words in Chapter 1.

KEY WORDS
constellation
stars
telescopes

1. A **constellation** is a group of _____. Select the best answer.
   A. stars   B. telescopes   C. planets   D. buildings

2. The ____constellation____ Ursa Major looks like a bear.

3. What do scientists use to look into space?
   Scientists use telescopes to look into space.

ORGANIZE IDEAS

As you read Chapter 1, complete the concept map.

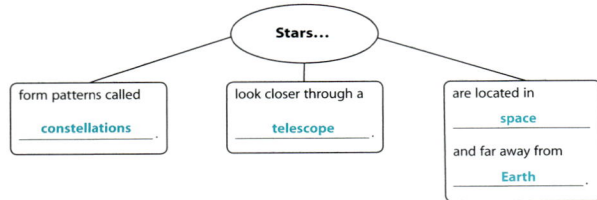

Stars...

form patterns called
constellations

look closer through a
telescope

are located in
space

and far away from
Earth

STRATEGY FOCUS: ASK QUESTIONS

What is one question you still have about stars?
Possible response: How many stars are there?

How could you find the answer to your question?
Possible response: I could look in my science book.

© 2008 MILLMARK EDUCATION

Stars and Galaxies: Our Universe
Chapter 1: The Sun and Other Stars

Study Guide
Learning Master 6

GREEN LEVEL
Student Book,
pages 4–8

USE KEY WORDS

Look at the Key Words on page 23 of your book.
Answer these questions about the Key Words in Chapter 1.

KEY WORDS
light-year
star

1. A **light-year** is a measure of _____.
   Select the best answer.
   A. brightness   **B. distance**   C. force   D. time

2. The **star** that is closest to Earth is ____the sun____

ORGANIZE IDEAS

As you read Chapter 1, complete the concept map to explain how energy forms in the sun and reaches Earth.

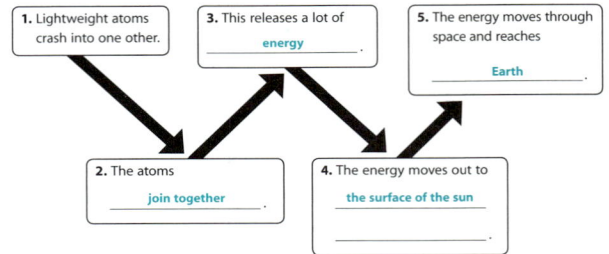

1. Lightweight atoms crash into one other.

3. This releases a lot of
   energy

5. The energy moves through space and reaches
   Earth

2. The atoms
   join together

4. The energy moves out to
   the surface of the sun

STRATEGY FOCUS: ASK QUESTIONS

After reading the chapter, what is one question you still have about stars?
Possible response: How hot are stars?

© 2008 MILLMARK EDUCATION

Stars and Galaxies: Changing Over Time
Chapter 1: A Look at Stars

Study Guide
Learning Master 7

ORANGE LEVEL
Student Book,
pages 4–8

USE KEY WORDS

Look at the Key Words on page 23 of your book.
Answer these questions about the Key Words in Chapter 1.
Select the best answer.

KEY WORDS
absolute magnitude
apparent magnitude
fusion
light-year
star

1. Which describes how bright a **star** looks?
   A. fusion   C. absolute magnitude
   B. light-year   **D. apparent magnitude**

2. Which describes how bright a **star** really is?
   A. fusion   **C. absolute magnitude**
   B. light-year   D. apparent magnitude

3. Which describes the combining of smaller atoms into one larger atom?
   A. fusion   C. absolute magnitude
   B. light-year   D. apparent magnitude

ORGANIZE IDEAS

As you read Chapter 1, complete the chart.

PROPERTIES OF STARS		
Property	**Definition**	**Example**
temperature	a measure of how hot or cold something is	Possible response: The temperature of a blue star is very hot.
brightness: apparent magnitude	how bright a star looks from Earth	A small, cooler star looks brighter from Earth than other stars because it is closer.
brightness: **absolute magnitude**	how bright a star really is	Possible response: A large, hot star is brighter than a small, cooler star.

STRATEGY FOCUS: ASK QUESTIONS

What is one question you had that was not answered in this chapter?
Possible response: How long would it take a spaceship to reach Proxima Centauri?

© 2008 MILLMARK EDUCATION

Stars and Galaxies: Exploring with Technology
Chapter 1: Collecting Energy from Space

Study Guide
Learning Master 8

PURPLE LEVEL
Student Book,
pages 4–8

USE KEY WORDS

Look at the Key Words on page 23 of your book.
Answer these questions about the Key Words in Chapter 1.

KEY WORDS
fusion
star

1. The combining of atoms is called ____fusion____.

2. A **star** shines because of ____fusion____.

3. What is a **star**?
   A star is a glowing ball of hot gases.

ORGANIZE IDEAS

As you read Chapter 1, complete the word map.

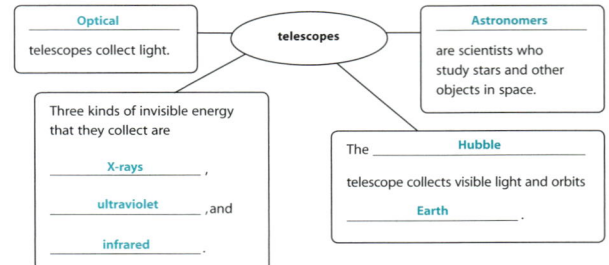

Optical
telescopes collect light.

telescopes

Astronomers
are scientists who study stars and other objects in space.

Three kinds of invisible energy that they collect are
X-rays,
ultraviolet, and
infrared.

The ____Hubble____ telescope collects visible light and orbits
Earth.

STRATEGY FOCUS: ASK QUESTIONS

What is one question you still have about telescopes?
Possible response: Where is the biggest telescope on Earth?

Where could you look to find the answer to your question?
Possible response: I could look on the Internet.

© 2008 MILLMARK EDUCATION

---

84   Chapter 1 Learning Masters: Answers

Learning Master 9

NAME _____

Stars and Galaxies: Light from Far Away
Chapter 2: Distances to Stars

Study Guide
Learning Master 9

BLUE LEVEL
Student Book,
pages 10-14

USE KEY WORDS

Look at the Key Words on page 23 of your book.
Answer these questions about Key Words in Chapter 2.

KEY WORDS
constellation
light-year
star

1. The sun is the closest ____star____ to Earth.

2. A group of **stars** makes a pattern called a ____constellation____.

3. A ____light-year____ is the distance light travels in one year.

ORGANIZE IDEAS

As you read Chapter 2, fill in the concept map with facts about stars.
Possible responses below

- Many stars are bigger than the sun.
- Stars are glowing balls of hot gases.
- Our closest star is the sun.
- **stars**
- Scientists measure distances to stars in light-years.
- Stars look small because they are so far away.
- Stars make their own light.

STRATEGY FOCUS: ASK QUESTIONS

What is one question you were able to answer while reading this chapter? What is the answer?

Possible response: What makes up a star? A star is made of hot gases.

Stars and Galaxies **35**

Learning Master 10

NAME _____

Stars and Galaxies: Our Universe
Chapter 2: Different Kinds of Stars

Study Guide
Learning Master 10

GREEN LEVEL
Student Book,
pages 10-14

USE KEY WORDS

Look at the Key Words on page 23 of your book.
Answer these questions about the Key Words in Chapter 2.
Select the best answer.

KEY WORDS
absolute magnitude
apparent magnitude

1. What does **absolute magnitude** describe?
   - **A.** how bright a star really is
   - B. how bright a star looks
   - C. how bright a star looks
   - D. how close a star looks

2. The closer a star is to Earth, the ____higher____ (higher/lower) the star's **apparent magnitude**.

ORGANIZE IDEAS

As you read Chapter 2, complete the chart. First, read the cause. Then, write the effect.
Possible responses below

Cause	Effect
1. Hotter stars have more atoms combining in them.	1. **More energy is released.**
2. Stars have different temperatures.	2. **Stars are different colors.**
3. A star with high absolute magnitude is far away from Earth.	3. The star looks **dim** from Earth and has a **low** apparent magnitude.
4. A star with low absolute magnitude is very close to Earth.	4. The star looks **bright** from Earth and has a **high** apparent magnitude.

STRATEGY FOCUS: ASK QUESTIONS

What new question do you have about the chapter?

Possible response: Will the sun ever become a red star?

Stars and Galaxies **37**

Learning Master 11

NAME _____

Stars and Galaxies: Changing Over Time
Chapter 2: The Life Cycle of a Star

Study Guide
Learning Master 11

ORANGE LEVEL
Student Book,
pages 10-14

USE KEY WORDS

Look at the Key Words on page 23 of your book.
Answer these questions about the Key Words in Chapter 2.
Select the best answer.

KEY WORDS
black hole
gravity
mass
nebula
red giant
supernova

1. What is the name for a huge cloud of gas and dust in space?
   - A. black hole
   - **B. nebula**
   - C. supernova
   - D. red giant

2. What does a dying supergiant star become?
   - A. a black hole
   - B. a nebula
   - **C. a supernova**
   - D. a red giant

3. Which phrase describes a **black hole**?
   - **A. strong gravity**
   - B. an explosion
   - C. a cloud of gas and dust
   - D. a large, cool star

ORGANIZE IDEAS

As you read Chapter 2, complete the chart to compare the end of the life cycle of a star of average mass with a star of high mass.

What happens	Star of average mass	Star of high mass
forms in a _____	nebula	**nebula**
_____ the first time	**collapses**	collapses
expands to become a _____	**red giant**	**supergiant**
_____ the second time	**collapses**	**collapses**
expands to become a	**planetary nebula**	**supernova**
center or remainder collapses to become a _____	**white dwarf**	neutron star or **black hole**

STRATEGY FOCUS: ASK QUESTIONS

What is one question you were able to answer while reading this chapter? What is the answer?

Possible response: How long will the sun shine? It will shine for about 5 billion years because it is halfway through its 10-billion-year life cycle.

Stars and Galaxies **39**

Learning Master 12

NAME _____

Stars and Galaxies: Exploring with Technology
Chapter 2: Exploring Stars

Study Guide
Learning Master 12

PURPLE LEVEL
Student Book,
pages 10-14

USE KEY WORDS

Look at the Key Words on page 23 of your book.
Answer these questions about the Key Words in Chapter 2.
Select the best answer.

KEY WORDS
black hole
gravity
light-year
mass
neutron star
pulsar
red giant
supernova

1. A large, cool star that has expanded from a star of average mass is called a _____.
   - **A. red giant**
   - B. supernova
   - C. light-year
   - D. pulsar

2. Stars form when _____ pulls gas and dust together.
   - A. a black hole
   - B. a supernova
   - **C. gravity**
   - D. mass

3. A rapidly spinning **neutron star** that gives off bursts of radio waves is called a _____.
   - A. supernova
   - **B. pulsar**
   - C. black hole
   - D. red giant

ORGANIZE IDEAS

As you read Chapter 2, write the correct terms in the boxes to show the differences between the life cycles of a star with an average mass and a star with a high mass.

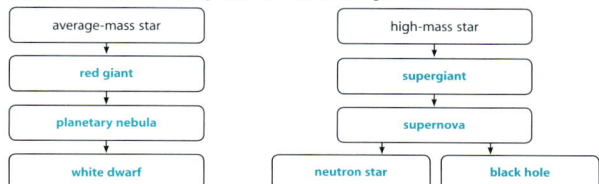

- average-mass star → **red giant** → **planetary nebula** → **white dwarf**
- high-mass star → **supergiant** → **supernova** → **neutron star** / **black hole**

STRATEGY FOCUS: ASK QUESTIONS

What is one question you still have about black holes?

Possible response: What happens to matter that gets pulled into a black hole?

Stars and Galaxies **41**

Chapter 2 Learning Masters: Answers 85

Learning Master 13

Stars and Galaxies: Light from Far Away
Chapter 3: Groups of Stars

BLUE LEVEL
Student Book,
pages 16–18

USE KEY WORDS

Look at the Key Words on page 23 of your book.
Answer these questions about the Key Words in Chapter 3.

KEY WORDS
galaxy
gravity
Milky Way Galaxy
stars

1. A **galaxy** is a group of ____ stars ____, gas, and dust.

2. **Stars** you can see with just your eyes are in the ____ Milky Way Galaxy ____.

3. How does **gravity** help form a **galaxy**?
   Gravity pulls together the parts of a galaxy.

ORGANIZE IDEAS

As you read Chapter 3, answer each question in the concept map.

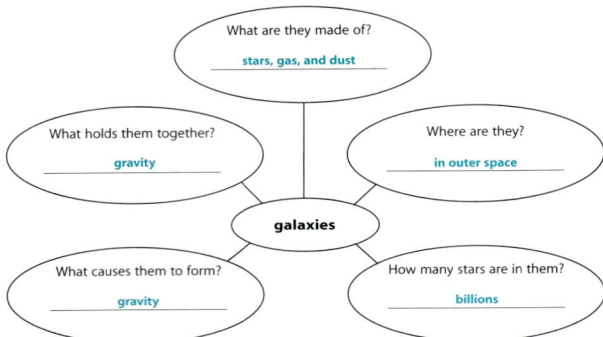

- What are they made of?
  stars, gas, and dust
- What holds them together?
  gravity
- Where are they?
  in outer space
- galaxies
- What causes them to form?
  gravity
- How many stars are in them?
  billions

STRATEGY FOCUS: ASK QUESTIONS

What is one question you still have about galaxies?

Possible response: How far away is the farthest galaxy we know about?

Stars and Galaxies **45**

© 2008 MILLMARK EDUCATION

Learning Master 14

Stars and Galaxies: Our Universe
Chapter 3: A Universe of Galaxies

GREEN LEVEL
Student Book,
pages 16–18

USE KEY WORDS

Look at the Key Words on page 23 of your book.
Answer these questions about the Key Words in Chapter 3.

KEY WORDS
galaxy
gravity
Milky Way Galaxy
telescopes
universe

1. A ____ galaxy ____ forms when **gravity** pulls stars together.

2. The **Milky Way Galaxy** is part of the ____ universe ____.

3. What do scientists use **telescopes** to study?
   They use them to study distant stars and galaxies.

ORGANIZE IDEAS

As you read Chapter 3, complete the concept map. In each box, write a science word that is connected to galaxies. It does not have to be a Key Word. Then write a definition for the word.

Possible responses below

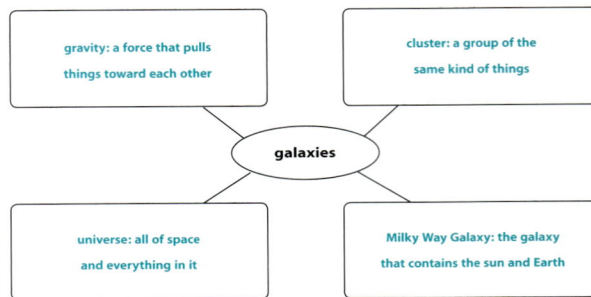

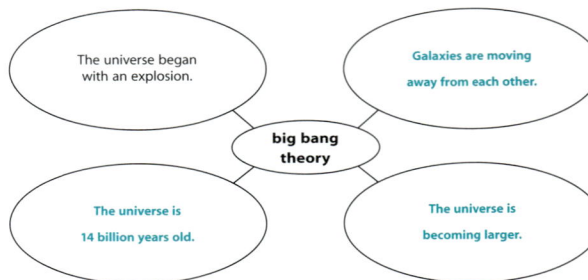

- gravity: a force that pulls things toward each other
- cluster: a group of the same kind of things
- galaxies
- universe: all of space and everything in it
- Milky Way Galaxy: the galaxy that contains the sun and Earth

STRATEGY FOCUS: ASK QUESTIONS

Write one question you have after reading Chapter 3.

Possible response: How does a telescope collect light from stars?

Stars and Galaxies **47**

© 2008 MILLMARK EDUCATION

Learning Master 15

Stars and Galaxies: Changing Over Time
Chapter 3: Groups of Stars

ORANGE LEVEL
Student Book,
pages 16–18

USE KEY WORDS

Look at the Key Words on page 23 of your book.
Answer these questions about the Key Words in Chapter 3.

KEY WORDS
galaxies
gravity
universe

1. Groups of stars, gas, and dust are called ____ galaxies ____.

2. ____ Gravity ____ holds **galaxies** together.

3. The ____ universe ____ contains all the **galaxies** in space.

ORGANIZE IDEAS

As you read Chapter 3, complete the chart by writing details about each type of galaxy.
Possible responses below

TYPES OF GALAXIES		
Spiral Galaxy	**Elliptical Galaxy**	**Irregular Galaxy**
1. It is shaped like a pinwheel.	1. It is shaped like a flattened ball.	1. It is not a spiral.
2. New stars form in the arms.	2. Some are round.	2. It is not elliptical.
3. The center is made of older stars.	3. About 60 percent of all galaxies are elliptical.	3. It is made up of mostly young stars.

STRATEGY FOCUS: ASK QUESTIONS

What is one question you still have about galaxies?

Possible response: What is the largest galaxy we know of?

Where could you look to find the answer to your question?

Possible response: I could look on the Internet or call a scientist at a university.

Stars and Galaxies **49**

© 2008 MILLMARK EDUCATION

Learning Master 16

Stars and Galaxies: Exploring with Technology
Chapter 3: Exploring the Universe

PURPLE LEVEL
Student Book,
pages 16–18

USE KEY WORDS

Look at the Key Words on page 23 of your book.
Answer these questions about the Key Words in Chapter 3.

KEY WORDS
big bang theory
galaxies
universe

1. The ____ big bang theory ____ is an idea that explains how the **universe** began.

2. Billions of ____ galaxies ____ exist in the **universe**.

ORGANIZE IDEAS

As you read Chapter 3, write science facts about the big bang theory.
Possible responses below

- The universe began with an explosion.
- Galaxies are moving away from each other.
- big bang theory
- The universe is 14 billion years old.
- The universe is becoming larger.

STRATEGY FOCUS: ASK QUESTIONS

What is a good source of information to answer any remaining questions you have? Explain why it would be a good source.

ANSWERS WILL VARY. Students might mention the Internet, encyclopedias, or textbooks. Students should give a reason for the source they selected, such as it comes from a famous university or a government agency.

Stars and Galaxies **51**

© 2008 MILLMARK EDUCATION

Stars and Galaxies: Light from Far Away
End-of-Book Test

MULTIPLE CHOICE

Read each item. Select the best answer.   1 POINT EACH

1. A pattern of stars is called a _____

   A. group   B. light-year   **C. constellation**   D. galaxy

2. Which is the closest star to Earth?

   A. Barnard's Star   B. Sirius   C. Proxima Centauri   **D. the sun**

3. Which object reflects light?

   A. the sun   B. a star   C. a light-year   **D. the moon**

4. Which force causes galaxies to form?

   A. dust   **B. gravity**   C. stars   D. gases

YES OR NO

Read each item. Write **YES** if it is a star. Write **NO** if it is not a star.   1 POINT EACH

5. the sun _____ **YES**

6. Earth _____ **NO**

7. the moon _____ **NO**

8. Sirius _____ **YES**

COMPLETE THE SENTENCE

Circle the correct word to complete each sentence.   1 POINT EACH

9. Scientists measure long distances in space in (**light-years**, constellations).

10. Every star belongs to a (constellation, **galaxy**).

Stars and Galaxies **57**

Stars and Galaxies: Our Universe
End-of-Book Test

MULTIPLE CHOICE

Read each item. Select the best answer.   1 POINT EACH

1. Which makes its own light?

   A. a planet   **B. a star**   C. Earth   D. the moon

2. The hottest stars are _____

   A. blue   B. yellow   C. red   D. white

3. The shape of the Milky Way Galaxy is _____.

   A. elliptical   B. round   C. irregular   **D. spiral**

FILL IN THE BLANK

Read the items. Write the missing words in the blanks.   1 POINT EACH

4. Scientists use _____ **telescopes** _____ to collect light from stars.

5. _____ **Absolute magnitude** _____ is how bright a star really is.

6. The _____ **universe** _____ has billions of galaxies.

7. _____ **Atoms** _____ in a star combine and release energy.

8. A small star might have a high _____ **apparent magnitude** _____ because it is close to Earth.

SHORT ANSWER

Read each question. Write an answer in 1-2 complete sentences.   1 POINT EACH

9. Why do scientists use light-years to measure distances to most stars?

   Scientists use light-years because most stars are so far away.

10. Compared to other stars, how big is the sun?

    Compared to other stars, the sun is a medium-sized star.

58 *Stars and Galaxies*

Stars and Galaxies: Changing Over Time
End-of-Book Test

MULTIPLE CHOICE

Read each item. Select the best answer.   1 POINT EACH

1. Which causes a star to give off energy?

   A. a nebula   B. gravity   **C. fusion**   D. a light-year

2. How long a star lives depends on its _____

   A. nebula   B. brightness   C. color   **D. mass**

3. Which is part of the life cycle of a high-mass star?

   A. a white dwarf   C. a red giant
   B. a supernova   D. a planetary nebula

4. What color star is the sun?

   A. white   B. orange   C. red   **D. yellow**

5. A galaxy that has the shape of a flattened ball is _____ galaxy.

   A. a cluster   B. an irregular   **C. an elliptical**   D. a spiral

6. Our solar system includes everything that moves around the _____.

   A. galaxy   **B. sun**   C. cluster   D. moon

SHORT ANSWER

Read each question. Write each answer in 2-3 sentences.   2 POINTS EACH

7. What is the difference between apparent magnitude and absolute magnitude?

   Apparent magnitude is how bright a star looks from Earth. Absolute magnitude is how bright a star really is.

8. How does a star form inside a nebula? Use the words **gravity** and **fusion** in your description.

   Some places in a nebula have a lot of gas and dust. Gravity pulls the gas and dust together into a ball. As more material gathers, the ball gets hotter. Fusion begins and a star is born.

Stars and Galaxies **59**

Stars and Galaxies: Exploring with Technology
End-of-Book Test

MULTIPLE CHOICE

Read each item. Select the best answer.   1 POINT EACH

1. Different kinds of telescopes collect different kinds of _____.

   A. light   B. X-rays   **C. energy**   D. atoms

2. The real brightness of a star depends on its _____.

   A. temperature and distance from Earth   C. size and distance from Earth
   B. size and temperature   D. distance from other stars

3. Which describes an invisible area where gravity pulls in all energy and matter?

   A. a mass   B. a neutron star   **C. a black hole**   D. a supernova

ORDER THE EVENTS

Number the events in the order they happen in the life cycle of a star of high mass. Write **1** for the first event, **2** for the second event, **3** for the third event, and **4** for the last event.   1 POINT EACH

4. _**4**_ Material becomes a black hole.

5. _**2**_ The star expands into a supergiant.

6. _**1**_ Gravity pulls gas and dust in a nebula into a ball.

7. _**3**_ The star explodes in a supernova.

SHORT ANSWER

Read the item. Write an answer in 2-3 complete sentences.   3 POINTS

8. Explain the big bang theory.

   The big bang theory states that the universe began about 14 billion years ago with an explosion. It blasted matter in all directions. The universe has been expanding ever since.

60 *Stars and Galaxies*

Assessment Learning Masters: Answers **87**

NAME _____

Ask Questions

Solar Storms

Strategy Assessment
Learning Master 23

ALL LEVELS

Storms on Earth can be rainy, snowy, or windy. But what is a storm like on the sun? A **solar storm** is a huge explosion from the outer part of the sun. A large cloud of gas blasts into space. This **solar flare** contains very hot particles.

Some of these particles reach Earth. We can't see the particles, but we can notice their effects. The particles can damage weather and communications satellites. They can even cause power blackouts.

Near the North and South Poles, the gas particles enter Earth's atmosphere. They collide with gases in the air and cause the sky to glow. These **auroras** happen regularly. But during a solar storm, the auroras are much brighter. They may look like white, green, red, and purple ribbons glowing in the sky.

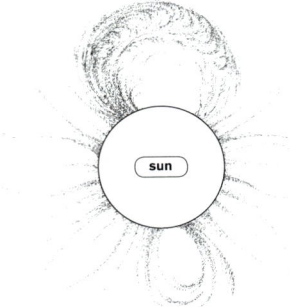

▲ The cloud of particles from a solar storm can grow to be as wide as 48 million kilometers (about 30 million miles) across. This is wider than the sun.

ASSESS THE COMPREHENSION STRATEGY

1. Look at the article title and diagram before you read the article. Write one question you have about solar storms. As you read the article, write another question. Then write one question you still have after reading the article.

BEFORE READING	DURING READING	AFTER READING
ANSWERS WILL VARY. Students' questions should show they are thinking about the information in the article before, during,	and after reading. Note whether they are able to answer any of the questions they ask. Possible responses	include What is a solar storm? What causes a solar storm? Do solar storms affect Earth?

2. After reading the article, reread your questions. Were you able to answer any of the questions? Write any answers you found. How can you find answers to the questions you still have?

Possible response: A solar storm is a huge explosion in the sun's outer atmosphere. I can look on the Internet to learn more about solar storms.

Stars and Galaxies **61**

© 2008 MILLMARK EDUCATION

MILLMARK EDUCATION CORPORATION

Ericka Markman, President and CEO; Karen Peratt, VP, Editorial Director; Lisa Bingen, VP, Marketing; Rachel L. Moir, Director, Operations and Production; Shelby Alinsky, Assistant Editor; Frances Jenkins, Science Editor; Kris Hanneman, Photo Research

PROGRAM AUTHORS

Mary Hawley; Program Author, Instructional Design
Kate Boehm Jerome; Program Author, Science

BOOK DESIGN

Steve Curtis Design

PROGRAM ADVISORS

Scott K. Baker, PhD, Pacific Institutes for Research, Eugene, OR
Carla C. Johnson, EdD, University of Toledo, Toledo, OH
Donna Ogle, EdD, National-Louis University, Chicago, IL
Betty Ansin Smallwood, PhD, Center for Applied Linguistics, Washington, DC
Gail Thompson, PhD, Claremont Graduate University, Claremont, CA
Emma Violand-Sánchez, EdD, Arlington Public Schools, Arlington, VA (retired)

TECHNOLOGY

Arleen Nakama, Project Manager
Audio CDs: Heartworks International, Inc.
CD-ROMs: Cannery Agency

IMAGE CREDITS, TEACHER'S GUIDE

cover ©Peter Arnold, Inc./Alamy
Illustrations: pgs. 15 and 53, Sharon and Joel Harris; pgs. 21, 61, 72a, 72b, 72c, 72d, 73, 75a, 75b, 75c, 75d, 75e, 75f, Greg Harris

PHOTO CREDITS, CONCEPT CONNECTOR

SIDE A: Background image ©Jerry Schad/Photo Researchers, Inc.; spiral galaxy and sun ©EUROPEAN SPACE AGENCY/Photo Researchers, Inc.; moon ©John Chumack/Photo Researchers, Inc.; Merik ©JTB Photo/Photolibrary; telescope ©NASA

Published by Millmark Education Corporation
7272 Wisconsin Avenue, Suite 300
Bethesda, MD 20814

ISBN-13: 978-1-4334-0234-0

Printed in the USA

10 9 8 7 6 5 4 3 2 1